DOMESTIC WATER TREATMENT

JAY H. LEHR
National Water Well Association

TYLER E. GASS
National Water Well Association

WAYNE A. PETTYJOHN
Ohio State University

JACK DeMARRE
Ecodyne-Lindsay Corporation

McGRAW-HILL BOOK COMPANY
New York St. Louis San Francisco Auckland Bogotá Hamburg Johannesburg London Madrid Mexico Montreal New Delhi Panama Paris São Paulo Singapore Sydney Tokyo Toronto

Library of Congress Cataloging in Publication Data
Main entry under title:

Domestic water treatment.

Includes bibliographical references and index.
1. Water—Purification. 2. Drinking water.
I. Lehr, Jay H., date.
TD433.D65 628.1'6 79-19941
ISBN 0-07-037068-0

Funded by National Water Well Association
in cooperation with U.S. Environmental Protection Agency
under Grant No. R804781-01

2 3 4 5 6 7 8 9 0 BPBP 8 9 8 7 6 5 4 3 2 1 0

The editors for this book were Jeremy Robinson and Ruth Weine,
the designer was Elliot Epstein, and the production supervisor
was Thomas G. Kowalczyk. It was set in Melior by Bi-Comp,
Incorporated.

Printed and Bound by The Book Press

Contents

Preface

This manual provides the first single-volume guide to all aspects of the analysis and treatment of water for domestic water-supply systems. It describes in detail both the techniques and the equipment available to produce high-quality water for residential use.

The book will prove invaluable as a ready reference for engineers and technicians, as an aid to government officials, and as a text for students. Homeowners, too, will find it very useful because it shows them how to meet the water-quality and -supply needs of their families. It also spotlights the latest types of water-conditioning equipment available for the home.

The first chapters provide a compact overview of the entire subject of water quality and treatment. Included are:

- A short summary of important facts about water chemistry so that the nonspecialist can follow subsequent descriptions of the various chemical processes that occur during water treatment
- Full descriptions of all pollution sources, both natural and of human origin
- An easy-to-follow explanation of the construction of surface and ground water supply systems for residential use

The main part of the book deals specifically with water sampling, analysis, and treatment techniques. It discusses the health hazards that result from certain elements, organisms, and sediment in a water-supply system. It describes in full detail the water-treatment techniques available to remove physical, biological, and chemical contaminants.

The book also covers water-quality requirements for special purposes (such as aquariums, house plants, household appliances, and the preparation of foods and beverages); the design, operation, and maintenance of water-treatment equipment; and the disposal of water-treatment wastes—a subject of particular concern to government agencies.

Domestic Water Treatment fills a long-standing need in the water-quality and -treatment field. It is the first book to cover this important subject thoroughly, yet in such clear terms that even the nonprofessional can benefit from it.

Jay H. Lehr

1

Introduction to Treatment of Individual Water Supplies

THE HYDROLOGIC CYCLE

Despite predictions to the contrary, the world is not running out of water. In fact, today's supply is likely to equal yesterday's and tomorrow's.

World water supply is maintained by the hydrologic cycle, a natural still generated by the sun (Fig. 1-1). Clouds are formed by water evaporated from the earth's surface and transpired by plants. As it cools, this atmospheric moisture condenses and replenishes water supplies as precipitation.

Some of this precipitation may not reach the ground. If it does, it will follow one of several routes: surface flow, to become streams; shallow ground infiltration to become soil moisture; or deep ground infiltration to become ground water. Approximately 70 percent of precipitation is returned to the atmosphere by evaporation and transpiration; 20 percent is runoff; and the rest becomes ground water.

The rate at which water enters the ground and returns to the atmosphere can vary according to soil conditions. At the beginning of a rain, water penetrates dry soil rapidly, but its flow decreases when a nearly constant rate of infiltration is reached. Frozen or saturated soil prevents water infiltration and promotes surface runoff, a common cause of floods.

If it rains long and hard enough, the soil will eventually become saturated and water will begin to drain into the earth through gravity until the soil reaches its field capacity. Water held in soil carries nutrients to plant tissues, which in turn yield water to the atmosphere through transpiration.

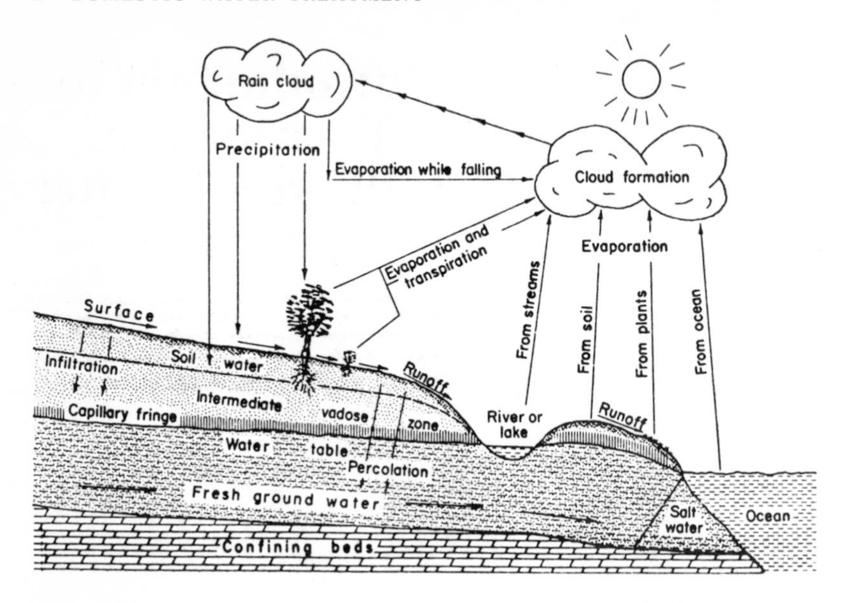

Figure 1-1. The hydrologic cycle.

HISTORY OF WATER TREATMENT

Long before the development of civilization, people recognized differences in water quality. Waters from various sources tasted sweet, salty, bitter, or sour; some were hot, while others were cold. Migrating tribes also knew that water at some oases, springs, or wells produced illness—usually diarrhea.

Water pollution has vexed us for centuries. Throughout its Middle East campaigns, Caesar's army had to boil water to ward off dysentery. Even earlier, Hippocrates had warned his colleagues to filter and boil water before drinking it. Some of the Dark Ages' worldwide plagues and epidemics were waterborne, since sanitary protection of water-supply facilities was practically nonexistent.

First Waterborne Diseases

Treatment of public water supply was held in minor regard until the end of the nineteenth century, when cholera epidemics in England showed that infected water supplies were responsible for public disease. The epidemic lasting from 1845 to 1849 claimed nearly 250 000 lives throughout Great Britain. In 1853, another cholera outbreak struck London, where nearly 11 000 perished. An investigation conducted in London's St. James parish traced the first 69 deaths to a public-supply well. A pipe draining a cesspool adjacent to the well had allowed infected waste to contaminate it.

In the United States, two major typhoid epidemics occurred 40 yr later. Schenectady, New York, which obtained its water supply from the Mohawk River, was the site of the first outbreak in July 1890. By October the disease had spread to Cohoes, a village situated at the confluence of the Mohawk and Hudson Rivers. A month later, the epidemic started downstream in West Troy. Albany, which also fronts the Hudson River, suffered a typhoid outbreak in December.

In 1891, the neighboring towns of Lawrence and Lowell, Massachusetts, suffered typhoid epidemics. Both communities obtained water from a surface supply contaminated by infected wastes. The New York and Massachusetts epidemics showed that rivers contaminated by sewage would transmit *Salmonella typhosa,* the agent of typhoid fever.

Early Water Treatment

Although drinking water should be palatable, it must be potable: free from disease or toxic substances. Disease-producing organisms (pathogens) will be transmitted by water contaminated by feces. In addition to causing epidemics of cholera and typhoid, waterborne pathogens have caused outbreaks of dysentery and infectious hepatitis. Most of these pathogens can be removed from water by filtration and chlorination.

The first large municipal water filtration system in the United States was built in Poughkeepsie, New York, in 1871. More sophisticated surface-water-treatment plants were constructed immediately before World War I, using chlorination to control outbreaks of typhoid fever. They were largely successful: between 1920 and 1930, typhoid mortality declined to one-sixth of the previous death rate.

Even though the threat of cholera and typhoid epidemics has been erased, outbreaks of gastrointestinal illness or taste and odor problems are still reported. Generally, these are due to toxic substances or dissolved minerals.

Toxic substances in water supplies may originate from a variety of sources: dissolved soil minerals, algae-generated phytotoxins, or metals dissolved from pipes and other pumping equipment. Industrial and household wastes, radioactive fallout, nuclear energy wastes, and pesticides may also generate toxic substances. Ingestion of these compounds spawns complex health problems, since many harmful substances are not removed during the routine water-treatment process.

Although most dissolved minerals are not toxic, they can produce unpleasant tastes or odors in water. Only within the past 40 to 50 yr have municipalities begun to remove them. In the 1930s, water-treatment plants began adding lime and soda ash to raw waters in order to decrease hardness. Oxidizing agents and filters were installed to reduce iron and manganese content, and activated carbon was used to control taste and odor.

Modern Water Treatment

Today's municipal water-treatment plants use a variety of processes, including screening, prechlorination, and sedimentation. Coagulation can remove sediment, turbidity, color, and organic matter; softening reduces hardness; and activated carbon removes objectionable tastes and odors. Clarification, filtration, and oxidation are all used to remove iron and manganese. Postchlorination is a routine final treatment process in most plants. Fluoride may be added to some supplies to protect children from tooth decay. However, most water-treatment plants do not use all these processes. There is still a wide range among final values of certain constituents in finished public water supplies (Table 1-1).

INDIVIDUAL WATER-SUPPLY TREATMENT

Aside from the development of primitive cistern filters and feeble attempts to improve well construction, few measures were taken to ensure safe private water supplies in the United States before 1920. Chlorination for pathogen control and the use of zeolites to soften water were early practices common before iron- and manganese-removal equipment became available. But it was not until the late 1940s that activated carbon was widely used in private water-supply systems to control taste and odor.

Synthetic detergents which alleviate some problems associated with water quality first appeared on the market in 1933. After World War II, these detergents gained widespread acceptance among homeowners, since they did not produce soap curd when used in hard water. Rather than removing offending elements from water, most detergents render them chemically inactive.

Modern water-conditioning equipment can solve almost all water-quality problems, but only a minority of the United States population actually uses it. Many individuals develop tolerances for drinking water of poor taste, odor, or appearance and believe that their water supply requires no treatment.

However, home treatment is often the only answer to many water-quality problems, particularly in rural and suburban areas. Also, many people using municipal water supplies supplement previous treatment with home water-conditioning units.

Table 1-1. Maximum, Median, and Minimum Values of Constituents and Properties of Finished Water in Public Water Supplies in the 100 Largest Cities in the United States, 1962*

Constituent or property	*Maximum*	*Median*	*Minimum*
Chemical analyses, ppm			
Silica (SiO_3)	72	7.1	0.0
Iron (Fe)	1.30	0.02	0.00
Manganese (Mn)	2.50	0.00	0.00
Calcium (Ca)	145	26	0.0
Magnesium (Mg)	120	6.25	0.0
Sodium (Na)	198	12	1.1
Potassium (K)	30	1.6	0.0
Bicarbonate (HCO_3)	380	46	0
Carbonate (CO_3)	26	0	0
Sulfate (SO_4)	572	26	0.0
Chloride (Cl)	540	13	0.0
Fluoride (F)	7.0	0.4	0.0
Nitrate (NO_3)	23	0.7	0.0
Dissolved solids	1580	186	22
Hardness as $CaCO_3$	738	90	0
Noncarbonate hardness as $CaCO_3$	446	34	0
Specific conductance, μ∫ at 25°C	1660	308	18
pH, pH units	10.5	7.5	5.0
Color, color units	24	2	0
Turbidity	13	0	0
Spectrographic analyses, μg/L			
Silver (Ag)	7.0	0.23	ND
Aluminum (Al)	1500	54	3.3
Boron (B)	590	31	2.5
Barium (Ba)	380	43	1.7
Chromium (Cr)	35	0.43	ND
Copper (Cu)	250	8.3	<0.61
Iron (Fe)	1700	43	1.9
Lithium (Li)	170	2.0	ND
Manganese (Mn)	1100	5.0	ND
Molybdenum (Mo)	68	1.4	ND
Nickel (Ni)	34	<2.7	ND
Lead (Pb)	62	3.7	ND
Rubidium (Rb)	67	1.05	ND
Strontium (Sr)	1200	110	2.2
Titanium (Ti)	49	<1.5	ND
Vanadium (V)	70	<4.3	ND
Radiochemical analyses			
Beta activity, pCi/L	130	7.2	<1.1
Radium (Ra), do	2.5	<0.1	<0.1
Uranium (U), μg/L	250	0.15	<0.1

* <: less than; ND: not detected.

Source: C. N. Durfor and Edith Becker, 1962, "Public Water Supplies of the 100 Largest Cities in the United States, U.S. Geological Survey Water-Supply Paper 1812.

Homeowners should be concerned about the quantity and the quality of their water supply. They should make sure that the water's chemical content is checked every 2 or 3 yr. Biological samples should be collected more frequently. If a water problem is discovered, it can usually be solved, as later chapters will show.

2

Introduction to Some Principles of Chemistry

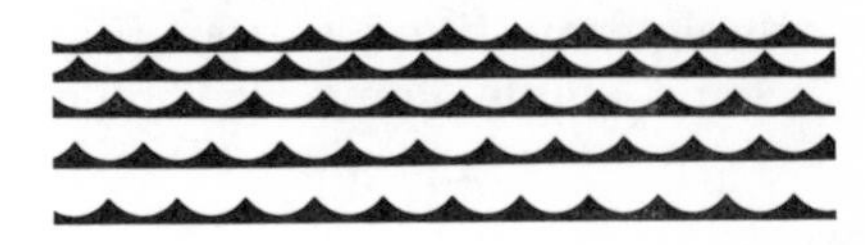

Water is a colorless, odorless, and tasteless substance. It exists naturally as either solid, liquid, or gas. Water is a very stable chemical compound: it can withstand temperatures up to 2700°C before decomposing into individual atoms. Here are some facts about water.

	Temperature Scale	
Property	*Celsius (Centigrade): °C*	*Fahrenheit: °F*
Boiling point	100	212
Freezing point	0	32

All matter can be classified as compounds, elements, or mixtures. To understand water chemistry, it is important to know how atoms are held together to form compounds.

Atoms

An atom is matter in its simplest form. Atoms consist of protons, neutrons, and electrons. Modern science has identified over 30 different kinds of atomic particles through the use of atom-smashing devices; but protons, neutrons, and electrons are still the most important. Protons are extremely small, positively charged particles; neutrons have no charge; and electrons are negatively charged. Compared with electrons, protons and neutrons are very large.

Hydrogen (H) is the smallest and simplest atom. It consists of one proton and one electron orbiting around it (Fig. 2-1). Other atoms have two, three, or more orbiting electrons; the most complex natural element, uranium, has 92. An atom always has as many protons as electrons, and it is electrically neutral.

Most atoms have a nucleus formed by neutrons and protons. Oxygen, for example, has eight protons and eight neutrons in its nucleus, with eight orbiting electrons (Fig. 2-2).

Some simple atoms are shown in Fig. 2-3.

Elements

The number of protons in an atom's nucleus determines which element it is. *Atomic number,* the number of protons in the nucleus, thus serves to identify elements.

Atomic mass units relate all atoms to each other by mass. These units were adopted to simplify measurement of the atom's extremely small particles (for example, electrons have a mass of 9.1×10^{-28} g).

Isotopes

Not all atoms of a particular element are identical. Atoms may vary in their atomic weight (the number of protons and neutrons in the nucleus). Since atoms must remain electrically neutral, their atomic weight can change only by adding neutrons. The number of protons always remains constant. Atoms of the same element with different atomic weights are called isotopes.

Hydrogen, for example, has three natural isotopes. The first, protium, has one proton and one electron. The second, deuterium, contains one proton, one electron, and one neutron. Tritium, hydrogen's third isotope, has one proton, two neutrons, and one electron. Remember, various isotopes are forms of the same element. They simply have a different number of neutrons.

Hydrogen's three isotopes are common in water. Tritium can be used

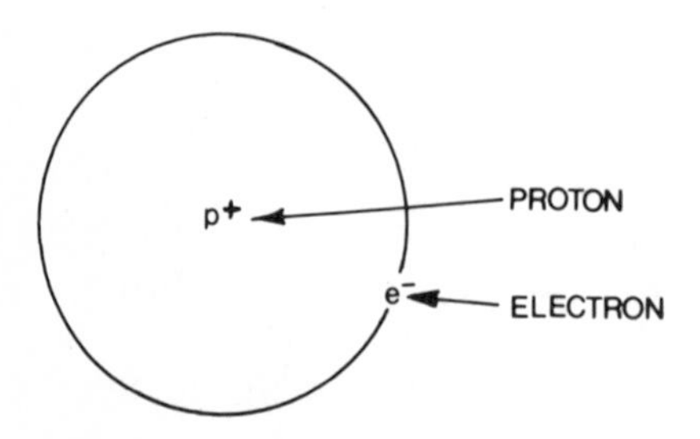

Figure 2-1. Hydrogen atom.

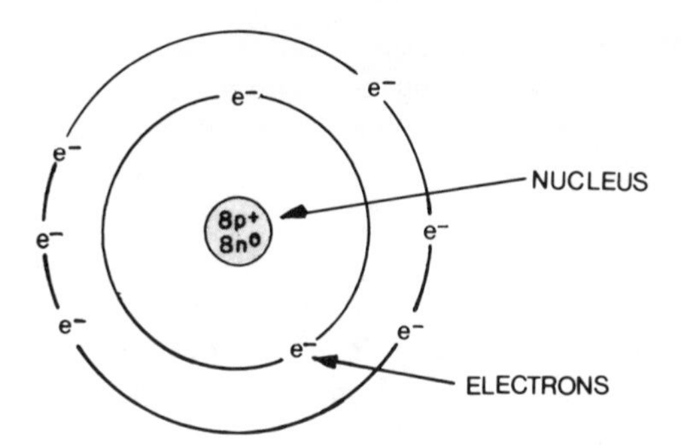

Figure 2-2. Oxygen atom.

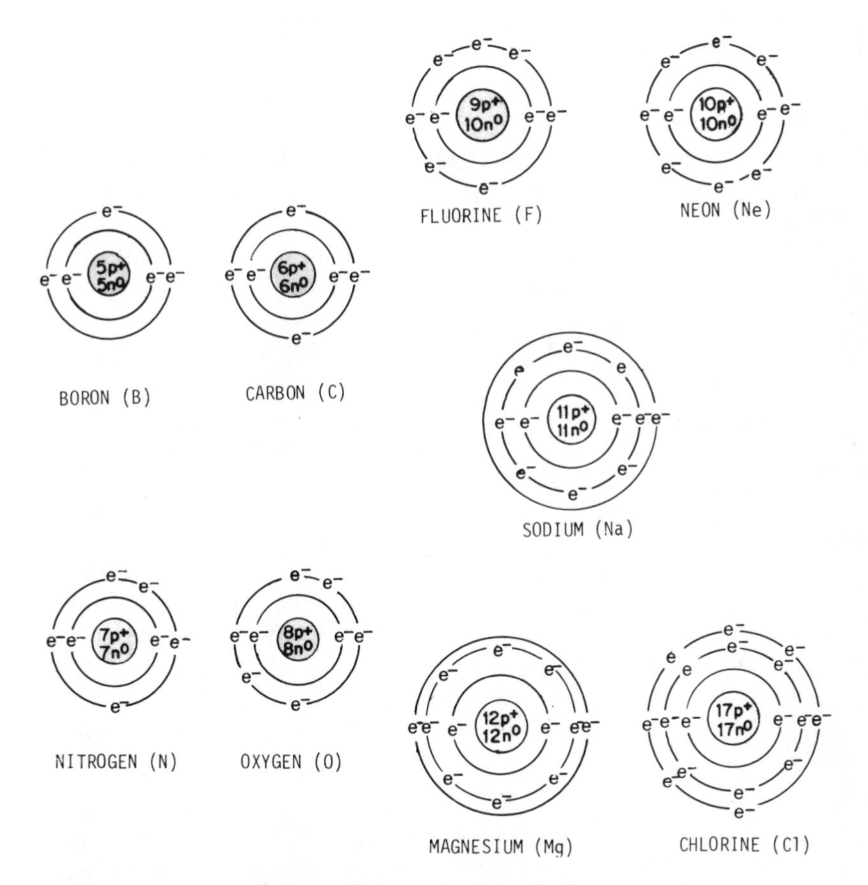

Figure 2-3. Atomic configurations of selected elements.

to determine how long ago a sample of ground water fell as rain. Since the amount of tritium in rain is constant, scientists know that some shallow well water in Illinois and Nebraska may have first fallen as rain some 50 to 100 yr ago.

Compounds

Most naturally occurring materials are compounds. A compound is a substance composed of two or more chemically bonded atoms or elements. The chemical properties of a given compound are different from those of its constituent elements.

Inert elements are those which may not combine with others. Their outermost orbit always contains eight electrons (except helium, which has two). These orbits are considered "full" (Fig. 2-3). Atoms with "unfilled" orbits can form compounds; those with "filled" orbits cannot. Helium and neon are examples of inert elements.

Valence

Atoms tend to form stable compounds by shifting their electronic structure. This is called valence bonding. An element's valence number tells how many of its electrons participate in forming a compound.

An atom's valence electrons are those occurring beyond the last filled orbit (free electrons). These may participate in a chemical reaction. To understand this, consider the formation of common table salt, or sodium chloride (NaCl). This compound is composed of the elements sodium (Na) and chlorine (Cl). Sodium has one electron in its outermost orbit, and chlorine has seven (Fig. 2-4). Sodium thus has an extra electron and chlorine lacks one. In a chemical reaction, sodium gives up its outer electron and develops a positive charge (1+). The chlorine atom accepts this electron in its outer orbit and develops a negative charge (1−). These charged atoms are called *ions*.

Some of the important elements in water chemistry are listed in Table 2-1. Several of them have more than one valence number. Chlorine, for example, can give up either one, three, five, or seven electrons, but it achieves its most stable structure when it gives up one.

Radicals

Some atoms called radicals group together and act as a single atom. One common radical is the negatively charged OH^- group. Other important water chemistry radicals are sulfate (SO_4^{2-}), nitrate (NO_3^-), ammonium (NH_4^+), and bicarbonate (HCO_3^-).

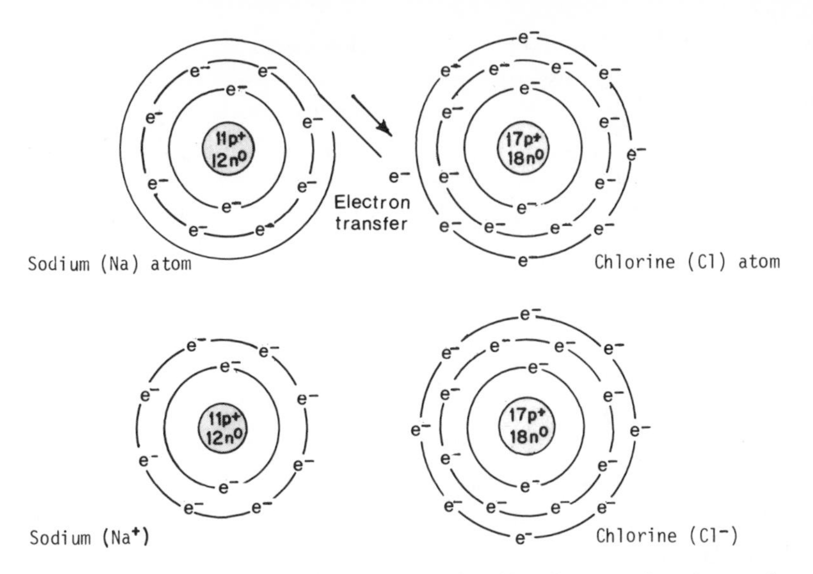

Figure 2-4. (Top) Sodium chloride is formed by the transfer of a single electron in the outer shell of the sodium atom to a vacancy that exists in the outer shell of the chlorine atom. (Bottom) The loss of an electron from the sodium atom leaves it positively charged; the gain of an electron in the chlorine atom makes it negatively charged. (*From Tracy, Tropp, and Friedl, 1970.*)

Chemical Bonding

Atoms are held together to form compounds by many types of chemical bonds. When atoms form compounds through electron transfer, they are held together by ionic bonds. Covalent bonding occurs when atoms "share" electrons. Water is an example of covalent bonding (Fig. 2-5). Oxygen needs two electrons to complete its outermost orbit, while hydrogen needs only one. By sharing electrons with two hydrogen atoms, oxygen can complete its outermost orbit (Fig. 2-5). Therefore, water's chemical formula is H_2O—two hydrogen atoms for each oxygen atom.

Common gases, such as oxygen, hydrogen, chlorine, and nitrogen, form molecules by sharing electrons. For example, since a hydrogen atom has only one electron in its outermost orbit, it can share that electron with another hydrogen atom. The symbol for these two atoms bonded covalently is H_2. Figure 2-6 shows how some of these simple molecules may appear.

Formulas

Chemical formulas tell how many and which kinds of atoms compose a compound. A chemical formula does not reveal which type of bonding

Table 2-1. Elements of Importance in Water Chemistry

Element	Chemical symbol	Atomic weight	Atomic number	Valence number	Corresponding equivalent weight
Aluminum	Al	26.98	13	3	8.99
Calcium	Ca	40.08	20	2	20.04
Carbon	C	12.01	6	2	6
				4	3
Chlorine	Cl	35.45	17	1	35.45
				3	11.82
				5	7.09
				7	5.06
Copper	Cu	63.54	29	1	63.54
				2	31.77
Fluorine	F	19.00	9	1	19.00
Hydrogen	H	1.008	1	1	1.008
Iron	Fe	55.85	26	2	27.03
				3	18.62
Magnesium	Mg	24.31	12	2	12.16
Manganese	Mn	54.99	25	2	27.49
				3	18.33
				4	13.75
				6	9.17
				7	7.86
Nitrogen	N	14.01	7	3	4.67
				5	2.80
Oxygen	O	16.00	8	2	8.00
Phosphorus	P	30.98	15	3	10.33
				5	6.20
Potassium	K	39.10	19	1	39.10
Silicon	Si	28.09	14	4	7.02
Sodium	Na	22.99	11	1	22.99
Sulfur	S	32.06	16	2	16.03
				4	8.02
				6	5.34

Source: New York State Department of Health.

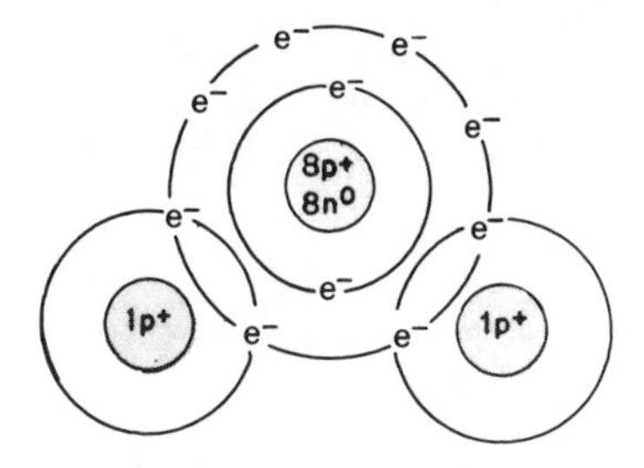

Figure 2-5. The formation of a molecule of water. Common electron-pairs are shared, completing shells of both elements. (*From Tracy, Tropp, and Friedl, 1970.*)

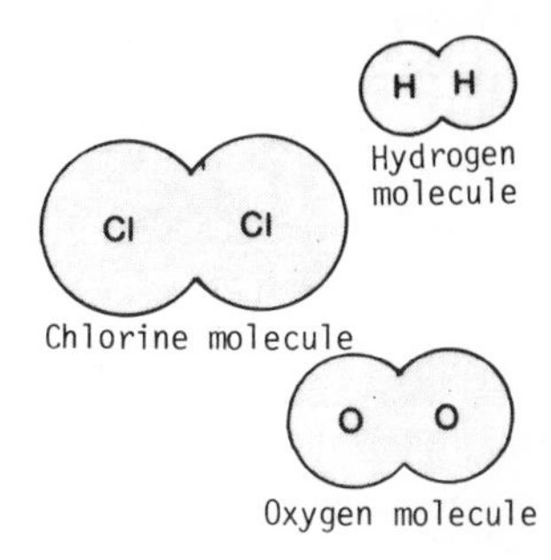

Figure 2-6. Laboratory models of molecules formed by electron sharing.

holds the atoms together. It is useful primarily in calculating formula weights and writing equations for chemical reactions.

Molecular Weight

A molecule is the smallest complete unit of a compound. The molecular weight of a compound equals the sum of the atomic weights of its constituent atoms. For example, to calculate the molecular weight of water, we add the weight of two hydrogen atoms (1.008 + 1.008) to the weight of an oxygen atom (16). The molecular weight of water is 18.016, or approximately 18.

Chemical Reactions

Chemical reactions occur when atoms combine to form new elements, or when elements break down into individual atoms. One common chemical reaction is the rusting of iron, or the chemical combination of iron and oxygen.

Chemical reactions are described mathematically in chemical equations. Writing chemical equations can be difficult, since they must be balanced. That is, the correct number and ratio of each atom involved in the chemical reaction must be calculated.

In the chemical equation for the rusting of iron, two iron atoms are required for every three oxygen atoms. Since oxygen exists only as a molecule (O_2), oxygen atoms may participate in a reaction only in pairs. Therefore, to maintain the ratio of 2 to 3, there must be six oxygen atoms for every four iron atoms. So the equation is written like this:

$$4Fe + 3O_2 \rightarrow 2Fe_2O_3$$

From a balanced equation, the weight of each element or molecule involved in the reaction can be calculated. Simply add the atomic weights of the various elements and molecules as given in Table 2-1.

$$4Fe + 3O_2 \rightarrow 2Fe_2O_3$$
$$4(55.85) + (3 \times 2 \times 16) = 2(2 \times 55.85 + 3 \times 16)$$
$$223.40 + 96 = 2(159.70)$$
$$319.40 = 319.40$$

These calculations show that 223.40 weight units of iron react with 96.00 weight units of oxygen to form 319.40 weight units of iron oxide. It does not matter whether the weight (or mass) units are grams, tons, pounds, etc., as long as they remain consistent.

Chemical Reactions in Water

Water's chemical nature causes the splitting of many molecules. Elements dissolve in water to become positively and negatively charged atoms, or ions. An example of ionization is

$$NaCl \rightarrow Na^+ + Cl^-$$

Positively charged atoms are called *cations;* negatively charged atoms are *anions.* In this example, sodium is the cation and chlorine the anion. When a molecule dissociates (ionizes) in water, the total charge before and after the reaction takes place must be the same. This condition is called electroneutrality.

Not all the molecules that ionize in water form individual atoms with positive and negative charges. Some become radicals with a net negative or positive charge. For example, calcium carbonate ($CaCO_3$) dissociates into Ca^{2+} and CO_3^{2-}. The calcium is a cation and the carbonate radical is an anion. CO_3^{2-} may later combine with free hydrogen in water to form HCO_3^-.

It should be mentioned that reactions do not always proceed from left to right. A reaction not in equilibrium will adjust itself under stress until equilibrium is reestablished.

This is called LeChâtelier's principle. Changing the temperature, pressure, or concentration of elements in a compound will upset its equilibrium and set off a chemical reaction. Water treatment uses this principle to change the chemical nature of a particular water supply.

Acids, bases, and salts form special ions when they dissolve in water. Acids yield hydrogen (H^+) cations. Hydrochloric acid dissociates as follows:

$$HCl \rightarrow H^+ + Cl^-$$

Strong acids such as hydrochloric (HCl), nitric (HNO_3), and sulfuric (H_2SO_4) acid yield an exceptionally large concentration of hydrogen ions. Weak acids, such as carbonic (H_2CO_3) and hydrogen sulfide (H_2S), ionize incompletely. Bases yield hydroxyl (OH^-) ions when dissolved in water. Bases, which can be either strong or weak, are also referred to as *alkalies*. (Basic substances are therefore alkaline.)

Since salts are composed of an acid and a base, their ionization is more complex. NaCl, a neutral salt, is neither acidic nor alkaline. It ionizes into Na^+ and Cl^-. Salts composed of strong acids and weak bases will first dissociate to form other acids. A second reaction then takes place, which yields hydrogen ions. An example of a reaction of this kind is

$$Al_2(SO_4)_3 \rightarrow 2Al^{3+} + 3(SO_4^{2-})$$

Since there are always free hydrogen (H^+) and hydroxyl (OH^-) ions in water, the second reaction of salts with water is called hydrolysis.

When salts composed of strong bases and weak acids dissolve in water, the reverse occurs. For example, calcium carbonate first yields carbonate ions. These ions combine with free hydrogen in water to form an excess of hydroxyl (OH^-) ions, producing a basic solution.

What happens when we combine equal weights of an acid with a base? When we add $HCl \rightleftharpoons H^+ + Cl^-$ (an acid) to $NaOH \rightleftharpoons Na^+ + OH^-$ (a base), the free H^+ and OH^- will combine to form water ($H_2O \rightleftharpoons H^+ + OH^-$ until an equilibrium of H_2O is achieved. The remaining Na^+ and Cl^- ions will be held in solution, unless the solution is already saturated.

This process (called neutralization) is important in water chemistry. Rain (an acid solution) is quickly neutralized as it reaches limestone in soil and bedrock. If rain were not neutralized by bases within the earth's soil, all water derived from precipitation would be acidic.

Another important characteristic of a solution is its pH, which indicates the relative concentration of hydrogen ions. Since concentration of hydrogen ions determines whether a substance is acidic or basic, the pH also gives a solution's relative acidity or alkalinity. If its pH is less than 7, a solution is acidic. Solutions with a pH greater than 7 are basic. Neutral solutions have a pH of 7.

Solutions

A solution consists of a solvent and a solute. A solvent promotes dissolving (dissolution); the solute is the substance being dissolved. Because water is the universal solvent, an almost infinite variety of solute substances and concentrations can exist in water.

A concentrated solution contains a large proportion of solute to solvent. A dilute solution has just the reverse. Most water chemistry is concerned with dilute solutions, and important solutions are expressed as the weight of solute per unit volume of solvent: e.g., milligrams per liter (mg/L). Most ions are measured in these units. Some, however, occur in such exceedingly dilute concentrations that they must be expressed in much smaller units—micrograms per liter (μg/L).

REFERENCES

New York State Department of Health: *Manual of Instruction for Water Treatment Plant Operators*, Health Education Service, Albany, N.Y.

Tracy, G., H. Tropp, and A. Friedl, 1970: *Modern Physical Science*, Holt, New York.

3
Sources of Water Pollution

INTRODUCTION

There is no such thing as "pure" water; all of it contains gases or minerals. Although these substances can be removed by treatment, the water may still retain some impurities and, in fact, the amount of dissolved minerals may be greater after the treatment than before (see (Chap. 9). Various techniques have been designed to remove unwanted substances from water, but the amount and type of substances removed depends on the treatment method. If one is willing to pay the price, all impurities can be extracted, but this is rarely, if ever, justified.

Water is polluted by both nature and the activities of human beings. In most cases, the pollution is hardly severe, and is not particularly detrimental to health. However, a few substances that are health hazards do occur in water, and can cause illness or even death. Other substances are merely undesirable, because they create bad tastes and odors, stain clothing and fixtures, or ultimately cost money. Still others have little or no effect in water used for most purposes.

NATURAL POLLUTION

Precipitation

"Pure as the driven snow" is a phrase that has been used with little question for decades. Only recently has the chemical quality of rain and snow been questioned, largely because of public awareness of the effects of air pollution. Examination of the data, however, reveals that the quality of precipitation, even before the evolution of human beings, was neither pure nor clean. Air pollution is not a phenomenon that developed in the twentieth century.

Moisture collects around minute particles in the atmosphere until a droplet forms and grows to such a size and weight that it falls. The nucleus of the droplet may consist of a particle of ice, dust blown from fields, ash emitted during volcanic eruptions, or crystals of salt evaporated from sea spray. Hazy summer afternoons occur because of the abundance of plant spores and pollens in the air which may also serve as raindrop nuclei. For millions of years the forests covering the land's surface have been ignited by lightning or spontaneous combustion. Their burning has released to the atmosphere gases and particulate matter—another source of natural air pollution.

The air also contains many gases. Carbon dioxide, formed largely from living plants and decaying vegetation, is an abundant atmospheric gas that combines with water to form carbonic acid. Decaying organic matter also produces hydrogen sulfide, which forms hydrosulfuric acid. Gases released from burning forests and grasslands also combine with moisture to form other acids. For these reasons rain has always been slightly acidic.

Surface Water

For millions of years, water in streams, lakes, and ponds has also been polluted by nature. Streams are turbid because of the silt and clay they transport as part of the earth's erosion cycle. Decaying algae and other organic matter may cause the water to have a dark color or an undesirable taste and odor; in fact, some algae release substances that are actually toxic. Every fall, leaves collect and deteriorate along the bottoms of lakes and streams, often making the water more acidic, creating a strong taste, and dyeing the water brown. Bacteria and protozoans also occur naturally in surface waters.

The flow of many streams comes in part from springs, some of which contain large concentrations of salts. The water may taste bitter, salty, or metallic; it may even taste and smell like rotten eggs. Streams also undergo an annual cycle of chemical content fluctuation. Water hardness in the Mississippi River, for example, doubles from summer to winter.

Ground Water Pollution

Water moves slowly through the ground, remaining in contact with soil and rocks for a long time. Therefore, various elements are dissolved from the enclosing soil and rocks, causing the water to increase in mineral content. For example, water that flows through limestone gains in hardness when calcium and magnesium ions are dissolved, but if the rocks consist of silt and clay (shale), sodium and potassium are released. Most rocks, and consequently much ground water, contain iron compounds that may impart a metallic taste and rusty color to water. These elements are natural pollutants.

Although some ground waters also contain acid and an abundance of trace elements, including poisonous arsenic, most of the chemicals in ground water are not particularly significant. In parts of western Texas, Colorado, and the Dakotas, for example, large concentrations of fluoride occur naturally in the ground water. The concentrations are so great that individuals who have consumed the water over a long period have developed mottled teeth, or (in a few cases) contracted fluorosis. On the other hand, when fluoride is present in optimal concentrations, it is of significant benefit because it reduces tooth decay.

Ground water may also come in contact with evaporite deposits such as gypsum and anhydrite. As it slowly dissolves these minerals, the water becomes bitter or salty. Certain kinds of bacteria also live in the ground and influence water quality, although in general none are pathogenic.

Without an actual chemical analysis of water from a particular well, it is simply not possible to determine the identity and concentration of the constituents naturally present in the water. A home, therefore, might not be furnished with the proper treatment equipment.

A general idea of the concentration of selected constituents that are present on a regional scale in water used for domestic and municipal purposes in the United States can be obtained from a report by Pettyjohn and others (1979). More detailed information can be obtained from the U.S. Geological Survey and from state and local water, soil, and health agencies.

It is evident that ground water can be polluted by natural events, but even though increased mineralization may create undesirable properties, in only rare cases does the water become toxic or detrimental to health. This is not the case, however, with many human-caused pollutants.

HUMAN-INDUCED POLLUTION

Pollution of Rain

Unlike ages long past, the present-day atmosphere contains a diversity of extraneous and commonplace materials. It steadily absorbs a wide range of solids, liquids, and gases resulting from human activities. These substances travel through the air, disperse, react among themselves and with other substances, both chemically and physically, and may spread to the four corners of the earth. They are washed out of the atmosphere by rain and snow.

Atmospheric contaminants in precipitated water include such things as radioactive fallout, pesticides attached to particles of clay, and toxic dust, such as chromium and asbestos. Gases generated by human beings may also combine with atmospheric water. These include lead, released by the burning of lead-base fuel, lead-arsenate spray residues, and the combustion by-products of coal. Chlorinated hydrocarbons are emitted from factory smoke stacks, from spraying, dust, and burning.

Perhaps today's best-known atmospheric contaminants are products of combustion. These include carbon monoxide, a poisonous gas which is almost exclusively manufactured; the acrid, corrosive, and poisonous sulfur dioxide from sulfur-rich fuels; and nitrous oxides, which under the influence of sunlight combine with gaseous hydrocarbons to form smog. Hydrocarbons and particulates are also products of combustion.

Surface-Water Pollution

Disposal of industrial, municipal, and domestic wastes directly into streams has been a major source of water pollution. In the Ohio River Valley, for example, the chemical industry periodically discharges wastes, including complex organic chemicals, into streams. In addition, the Ohio River and many other water courses are polluted by sulfuric acid draining from coal and metal mines (Biesecker and George, 1966). Acid drainage contributes large concentrations of sulfate, metals, hardness, and dissolved solids to ground and surface waters.

Villages and municipalities have disposed of their sewage directly into water courses for centuries. Sewage consumes much of a stream's dissolved oxygen and contains large concentrations of bacterial and viral forms, some of which may be pathogens. When municipalities process industrial wastes, many of the unusual chemicals which the wastes contain are not removed during waste-water treatment. As a result, the receiving water may contain practically no dissolved oxygen; it can be highly mineralized, particularly during periods of low flow; and it may have bad tastes and odors.

Even agricultural activities pollute surface waters. Evaporation consumes much of the water applied during irrigation, concentrating the salts that are present in the water and soil. The remaining water may either infiltrate, where it becomes more highly mineralized, or flow over the ground and into streams. The mineralized ground water may also reach nearby streams, contributing to the dissolved solids and hardness concentrations. Streams are also polluted by sediment eroded from agricultural lands and construction sites.

Pollution of Ground Water

Human sources and effects of ground water pollution are not widespread. In fact, substances naturally appearing in water supplies are far more widespread and significant to the general population than are human pollutants; all ground waters are subject to the rigors of nature whereas only a minute fraction of the supply has or can be adversely affected by human activities. The quality problems of human-influenced ground water are most commonly related to (1) water-soluble products that are placed on the land surface and in streams; (2) substances that are deposited or stored in the ground above the water table; and (3) disposal, storage, or extraction of material below the water table. Many of the pollution problems related to these situations are highly complex, and some are not yet well understood.

Because most individual home water-supply systems depend on wells, sources of ground water pollution are covered in considerable detail in this manual. They are particularly significant, since many of the reactions that tend to purify water in streams and lakes, such as dilution and biological or biochemical degradation, do not occur underground. Once ground water is polluted, it may require years or even decades before the resource returns to its normal state. Therefore, water treatment of individual well supplies may be necessary.

The effects of ground water pollution on a well-water supply may take a long time to appear as the contaminants slowly migrate toward the well and increase in concentration. In some situations contaminants may appear almost instantly. For these reasons a water-supply system should be checked periodically by chemical analysis, as described in Chap. 6. If the water shows, in particular, a dramatic change in taste, odor, or color, a state or local water or health agency should be contacted immediately, and water samples should be collected for analysis. When the sample is sent to a laboratory, it should be accompanied by a letter describing any activities, such as land filling, irrigation, waste disposal, road salting, oil-well drilling, or accidental spills which might have caused ground water pollution. This will enable the chemist to examine the sample for substances that might be characteristic of the waste. It is important to remember that some water pollutants only degrade water quality, while others may be toxic or hazardous in some other manner. Gasoline in water, for example, has caused many explosions and fires.

Ground water pollution problems exist, but they are not common. They do, however, create special problems. The major causes of ground water pollution are described in the following pages; some are obvious, but others are subtle. Remember, however, that the quantity of water stored in the ground is exceedingly large, and only a minute fraction of it has been polluted. Nevertheless, nearly all the existing polluted water is in areas that are inhabited.

Septic Tanks, Cesspools, and Privies

Probably a major cause of local ground water pollution in the United States is effluent from septic tanks, cesspools, and privies (Fig. 3-1). For the most part of little significance individually, these devices are important in the aggregate because they are so abundant and are found in every area not served by municipal or privately owned sewage-treatment systems. The area that each point source affects is generally small, since the quantity of effluent is small. In some limestone areas, however, the effluent may travel long distances in subterranean cavern systems with little change except dilution.

Most of the health problems that arise from this type of pollution are caused by recycling of sewage effluent through nearby water-supply systems. Such problems are especially common in areas where soils are thin and lie directly over fissured rock, such as some limestone, granite, and sandstone formations. In general, health hazards arise where soil characteristics are extreme—in tight, impermeable soils where proper infiltration cannot occur, and in loose, highly permeable conditions where soil lacks adequate adsorptive and filtration capacity to provide treatment of sewage.

Water supplies degraded by effluent are likely to contain abnormal concentrations of nitrate, chloride, sulfate, hardness, dissolved solids, detergents, and bacteria. The greatest danger in consuming untreated water polluted by sewage effluent is the potential for epidemics of waterborne diseases, such as typhoid fever, hepatitis, and the catchall illness, gastroenteritis.

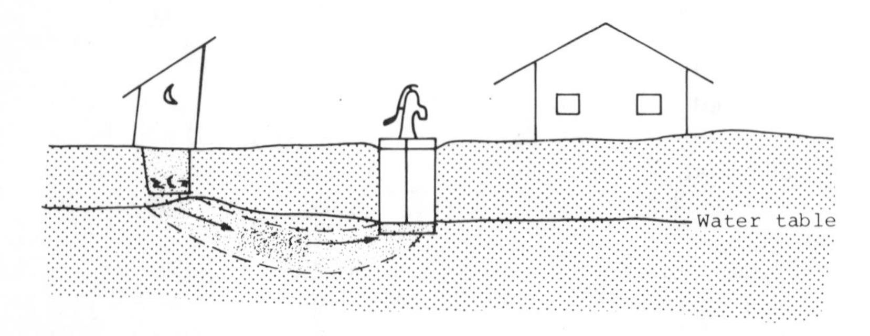

Figure 3-1. Shallow wells are commonly contaminated by effluent originating at cesspools and privies.

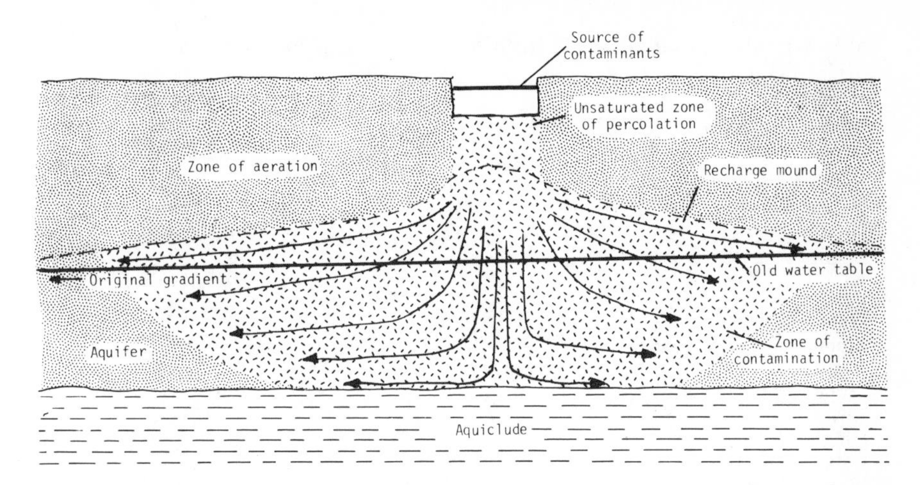

Figure 3-2. Percolation of contaminants through the zone of aeration and in an isotropic aquifer. (*From Deutsch, 1965.*)

Excavations

Excavations are adapted to a wide spectrum of municipal, industrial, and agricultural uses, including the storage, processing, treatment, and disposal of liquid and solid wastes. For example, lagoons are used to treat raw, secondary-, or tertiary-treated sewage; they also serve as spreading basins to dispose of effluent from treatment facilities and for ground water recharge to dispose of effluent. Unlined excavations are used by industry to pond cooling waters, to store waste water for later treatment or discharge, to evaporate or concentrate recoverable salts, and to dispose of wastes. In large agricultural operations, poultry and livestock wastes are commonly disposed of by spreading them over the ground. Excavations are not uncommon, though; liquid wastes from feedlots may be collected in ponds.

Although storage of municipal and agricultural wastes in pits, ponds, and lagoons has created local but significant ground water pollution problems, the most insidious effects come from industrial wastes. The number of industrial plants without waste-treatment facilities is exceptionally large.

Constituents found in typical municipal waste waters and their concentrations are shown in Table 3-1. Many of these substances infiltrate the sides and bottoms of holding structures (Fig. 3-2). The substances listed in Table 3-1 are similar to agricultural wastes, except for the lack of measurable concentrations of fertilizers and pesticides. Typical chemical characteristics of wastes from specific industrial sources are shown in Table 3-2.

Table 3-1. Municipal Waste-Water Characteristics (*Modified from Miller and Pettyjohn, 1974*)

	Concentration, mg/L (except as noted)	
Constituent	(1) *Untreated sewage*	(2) *Typical secondary-treatment effluent*
Physical		
Total solids	700	425
Total suspended solids	200	25
Chemical		
Total dissolved solids	500	400
pH, units	7.0 ± 0.5	7.0 ± 0.5
BOD	200	25
COD	500	70
Total nitrogen	40	20
Nitrate nitrogen	0	
Ammonia nitrogen	25	
Total phosphorus	10	10
Chlorides	50	45
Sulfate	—	
Alkalinity ($CaCO_3$)	100	
Boron	—	1.0
Sodium	—	50
Potassium	—	14
Calcium	—	24
Magnesium	—	17
Sodium adsorption ratio	—	2.7
Biological		
Coliform organisms, MPN/100 ml	10^6	

Sources: Column 1, Medium strength, Metcalf & Eddy, Inc. *Wastewater Engineering*, McGraw Hill, New York, 1972; column 2, CRREL Special Report 72-1.

In recent years the literature describing the extent of ground water pollution from excavations has become voluminous. Merely a few examples of pollution originating at holding or disposal ponds and lagoons are shown in Table 3-3.

It is evident that fluids commonly leak from excavations to contaminate ground water, rendering unsafe or undesirable local supplies of water that could be used for drinking. Furthermore, contaminated ground water flows to points of discharge, such as wells, springs, seeps, lakes, and streams. This leads to contamination of not only well water but of surface supplies as well. Only recently have investigators begun to realize the effect of contaminated ground water on surface-water supplies.

Alum Creek, a central Ohio stream, contains abnormally high concentrations of chloride due to ground water pollution brought about by

Table 3-2. Summary of Industrial-Waste Characteristics (*Modified from:* Newmerow, *Theory and Practice of Industrial Waste Treatment*, Addison-Wesley, Copyright 1963)

Industries producing wastes	*Major characteristics*
Food and Drugs	
Canned goods	High in suspended solids, colloidal and dissolved organic matter
Dairy products	High in dissolved organic matter, mainly protein, fat, and lactose
Brewed and distilled beverages	High in dissolved organic solids containing nitrogen and fermented starches or their products
Meat and poultry products	High in dissolved and suspended organic matter, blood, other proteins, and fats
Beet sugar	High in dissolved and suspended organic matter containing sugar and protein
Pharmaceutical products	High in suspended and dissolved organic matter, including vitamins
Yeast	High in solids (mainly organic) and BOD
Pickles	Variable pH; high in suspended solids, color, and organic matter
Coffee	High in BOD and suspended solids
Fish	Very high in BOD, total organic solids, and odor
Rice	High in BOD, total and suspended solids (mainly starch)
Soft drinks	High pH, suspended solids, and BOD
Apparel	
Textiles	Highly alkaline; colored; high BOD and temperature; high in suspended solids
Leather goods	High in total solids, hardness, salt, sulfides, chromium, pH, precipitated lime, and BOD
Laundry trades	High turbidity, alkalinity, and organic solids
Chemicals	
Acids	Low pH and organic content
Detergents	High in BOD and saponified soaps
Cornstarch	High in BOD and dissolved organic matter, mainly starch and related material
Explosives	TNT; colored, acidic odorous; contains organic acids and alcohol from powder and cotton, metal, acid, oils, and soaps
Insecticides	High in organic matter, benzene-ring structures; toxic to bacteria and fish; acidic
Phosphate and phosphorus	High in organic matter, benzene-ring structures; toxic to bacteria and fish; acidic
Formaldehyde	Normally has high BOD and HCHO; toxic to bacteria in high concentrations

Table 3-2. Summary of Industrial-Waste Characteristics (*Continued*)

Industries producing wastes	*Major characteristics*
Materials	
Pulp and paper	High or low pH, color; highly suspended colloidal and dissolved solids; inorganic fillers
Photographic products	Alkaline; contains various organic and inorganic reducing agents
Steel	Low pH, acids, cyanogen phenol, ore, coke, limestone, alkali, oils, mill scale; fine suspended solids
Metal-plated products	Acidic; metallic; toxic; low volume; mainly mineral matter
Iron-foundry products	High in suspended solids, mainly sand; some clay and coal
Oil	High in dissolved salts from field; high in BOD, odor, phenol, and sulfur compounds from refinery
Rubber	High in BOD, odor, suspended solids, chlorides; variable pH
Glass	Red color; alkaline; nonsettleable suspended solids
Naval stores	Acidic; high BOD
Energy	
Steam power	Hot; high volume; high in inorganic and dissolved solids
Coal processing	High in suspended solids; mainly coal; low pH; high H_2SO_4 and $FeSO_4$
Nuclear power and radioactive materials	Radioactive elements; can be very acidic and "hot"

the disposal of oil-field brines in ponds in the upper part of the basin. The contaminated ground water continuously flows into the stream; but stream-quality deterioration is most noticeable during summer and fall periods of low flow when ground water provides all or nearly all of the stream's discharge. This phenomenon has led to serious water-treatment and taste problems at the City of Westerville, which draws its supplies entirely from Alum Creek (Pettyjohn, 1975). This situation is not unique.

In addition to active containment structures, abandoned ponds and even ponds which have been filled with earth and reclaimed may be major sources of water pollution. A sample from an old, filled oil-field brine pond in central Ohio that had been reclaimed 5 yr previously showed concentrations of chloride near the water table to be in excess of 22 000 mg/L, but in adjacent unpolluted areas it was only 12 mg/L. Obviously, chloride will leach slowly from the reservoir for decades.

Accidental Spills of Hazardous Materials

Many kinds of toxic and hazardous materials are transported throughout the world by ship, truck, rail, and aircraft. Accidental spills are not uncommon, and virtually no foolproof methods are available to clean up a spill quickly and adequately. During a rail accident in central Ohio, three tank cars carrying thousands of gallons of organic chemicals ruptured and burned. Within 12 h, a nearby domestic well began to produce a foul odor and had to be abandoned immediately. Spills of atomic wastes constitute a special problem because they not only are highly toxic, but have half-lives of thousands of years.

A growing problem of substantial consequence is leakage from storage tanks and pipelines leading to such tanks. Gasoline leakage has caused severe problems throughout the nation. Gasoline floats on the

Table 3-3. Examples of Contamination from Holding Ponds

Type of waste	*Location*	*Contaminant*	*Areal extent*
Ammonium hydroxide	Oregon	SO_4, high hardness	>1 mi downgrade
Cd plating waste	New York	Cd, Cr	4300 × 100 ft
Oil-field brine	Arkansas	NaCl	4.5 mi^2
Radioactive	South Carolina	Sr^{90}	500 ft
Radioactive	Washington	Ru^{106}, Te^{99}	15 mi
Pesticides	Colorado	2, 4-D, As, F, CL	12 mi^2
Cu mine	Montana	Ca SO_4	
Kraft paper	Montana	Dissolved solids	
Sugar beets	Montana	BOD, odor	
Slaughterhouse	Montana	BOD, odor	
Mine tailings	Idaho	Pb, Zn, Cd	
Sewage	Washington	ABS, coliforms	200 ft +
TNT	Washington	Picric acid	
Wood chips	Oregon	Tannic acid	1500 × 500 ft
Pulp mill	Washington	Sulfite	
Municipal sewage	California	ABS	
Paper wastes	Wisconsin	BOD, TOC, odor, SO_4	3300 ft
Paper wastes	Wisconsin	BOD, TOC, odor, SO_4	0.75 mi^2
Milk whey	Wisconsin	Coliforms, odor, Cl	0.75 mi^2
Metal wastes	Wisconsin	Cl, F, SO_4, trace metals	
Wastes disposal company	Texas	Fe, Zn, Mn, Cd oils	
Wastes disposal company	Ohio	Cl, SO_4, F, heavy metals	0.5 mi^2
Oil-field brine	Ohio	Na Cl	
Plating wastes	Michigan	Cr	
Chemical wastes	West Virginia	Organics, Cl, SO_4	0.5 mi^2

ground water surface and leaks into basements, sewers, wells, and springs, causing noxious odors and tastes, explosions, and fires.

Literally thousands of miles of buried pipelines crisscross the United States. Leaks, of course, do occur, but it may be exceedingly difficult to detect and locate them. They are most likely to develop in transmission lines carrying corrosive fluids, though sometimes they result from faulty design or uneven settling of pipes and storage tanks.

Chromium compounds that polluted several shallow wells in Michigan were traced to a leaky sewer transporting metal-finishing wastes. The several leaks reported at the Hanford A.E.C. Works resulted from loaded, underground tanks which settled unevenly into the underlying earth materials, causing the pipelines carrying radioactive waste to break at joints.

Wastepiles and Stockpiles

Many waste materials are piled on the ground and abandoned or are stored to await final disposal. Nearly all wastepiles are exposed to precipitation, thus allowing the formation of a leachate that may flow over the land or infiltrate (Fig. 3-3). Some wastepiles not only pollute water, but produce offensive odors and provide a place for insects and other disease-carrying or disease-producing organisms to breed.

Stockpiling of raw materials or products may inadvertently lead to ground-water pollution. Many of these valuable resources, such as ore

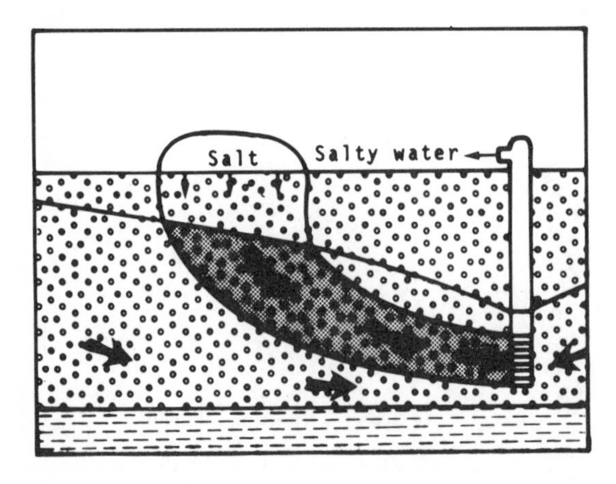

Figure 3-3. Leaching of solids at the land surface. The possibility of ground water pollution under these conditions is rarely anticipated.

and coal, contain materials that are water-soluble. Prime examples of stockpiles that have caused ground water pollution include salt for ice control, manure for fertilizer, and certain ores.

In recent years, particularly since the construction of the interstate highway system, water pollution due to road salting has become an increasing problem. In addition to killing trees and other vegetation, the salting brings about deterioration of stream quality due to highly mineralized surface runoff, and the infiltration of briney water causes ground water pollution.

Mining Activities

Mining is responsible for a wide variety of water-pollution problems. Some of the activities which create these problems are the pumping of mine water to the surface, the leaching of the spoil material, the natural discharge of waters from the mine, and the outpouring of milling wastes. Literally thousands of miles of streams and hundreds of acres of aquifers have been polluted by the highly corrosive mineralized waters from coal mines and dumps in Appalachia. In many western states, mill wastes and leachates have also seriously affected both surface and ground water.

Agricultural Activities

An increasing amount of both fertilizers and pesticides is being used in the United States each year. Many of these substances are highly toxic. In heavily fertilized areas, the infiltration of nitrate, a decomposition product of ammonia fertilizer, has frequently polluted ground water. The consumption of nitrate-rich water leads to the serious disease methemoglobinemia in infants.

In Colorado, automatic fertilizer feeders attached to irrigation-sprinkler systems are becoming increasingly popular. However, their operation may create a partial vacuum in the lines, causing fertilizer to flow from the feeder into the well. Even more serious is the suspicion that some individuals are dumping fertilizers directly into the well to be picked up by the pump and distributed to the sprinkler system. Such practices lead to direct and dangerous ground water pollution.

Animal feedlots cover relatively small areas but provide a huge volume of waste. These wastes have polluted both surface and ground water with large concentrations of nitrate. Even small feedlots and stables have created significant problems. In one situation (Pettyjohn, 1972), infiltration of livestock wastes continued to pollute a public well more than 40 yr after the livery stable was abandoned (Fig. 3–4).

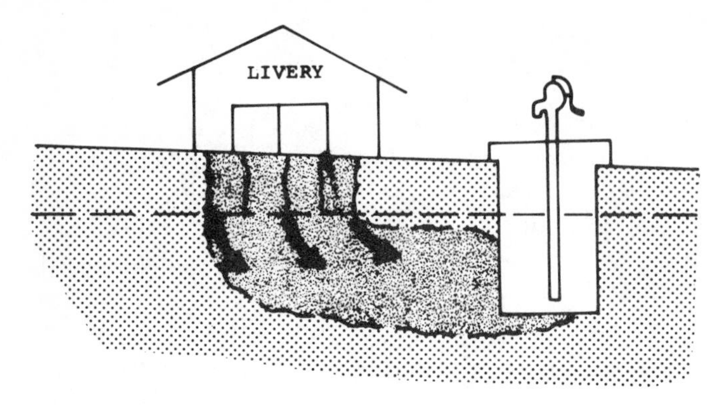

Figure 3-4. Many shallow wells have been contaminated by the infiltration of animal wastes.

Infiltration of Polluted Surface Water

The yield of many wells tapping streamside aquifers is sustained by infiltration of surface water (Fig. 3-5). In fact, more than half the well yield may be derived directly from induced recharge from a nearby stream, which may be polluted. As the induced water migrates through the ground, a few substances (although by no means all) are diluted or removed if the water flows through filtering materials, such as sand and gravel. Filtration is less likely to occur, however, if the water flows through large openings, such as those found in carbonate aquifers. Many pollutants, such as chloride, nitrate, and sulfate, are highly mobile, move freely with the water, and are not removed by filtration.

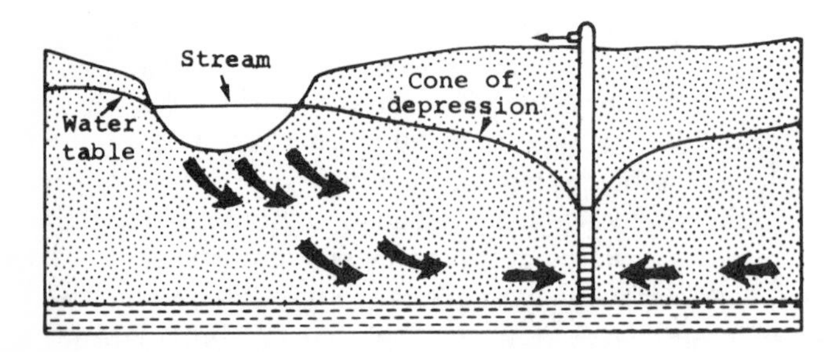

Figure 3-5. Cone of depression expanding beneath the river bed creates a hydraulic gradient between the aquifer and river. This can result in induced recharge to the aquifer from the river.

Landfills, Dumps, and Excavations

Sanitary landfills are generally considered to be those disposal landfills, or dumps, which are covered with soil daily. They are located as mounds, on top of the ground; beneath the surface of the ground, but above ground water; and in the ground below the ground water table. In many instances they are a potential hazard to ground water quality. Generally it has been the practice to dispose of toxic wastes, including pesticides, industrial by-products and residues, brines, acids, and other chemical compounds; municipal garbage; and inert or nearly inert insoluble substances, such as masonry demolition wastes, all in a common landfill. The only protection against the obvious hazard to health has been a result of fortuitous dilution—that is, the volume of toxic products has been low enough as compared with apparently nontoxic products to prevent any widespread pollution of the ground water body. When soluble materials in a landfill come into contact with water, the resultant leachate presents a threat to ground water quality (Fig. 3-6). The source of water within the landfill may be water that was buried with the material when disposed, as in the case of a municipal or industrial sludge, or ground water that is in contact with the buried refuse. Most commonly it is simply rainfall that enters the disposal site.

Following the cessation of various mining activities, the excavations are commonly abandoned (most states now require back-filling); eventually they may fill with water. These wet excavations have been used as dumps for both solid and liquid wastes. The wastes, in direct contact with an aquifer, may cause extensive pollution. Furthermore, highly concentrated leachates may be generated from the waste due to seasonal fluctuations of the water table.

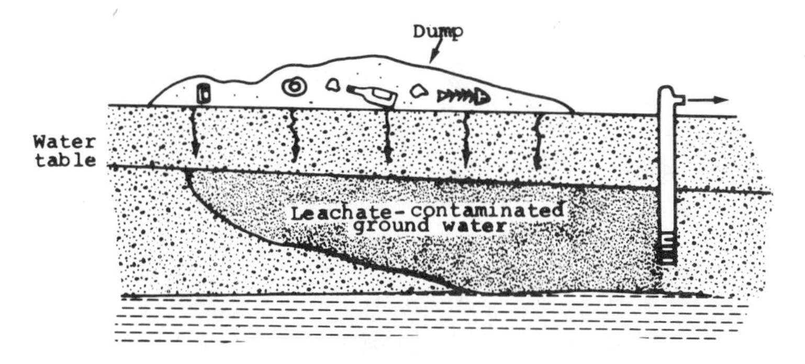

Figure 3-6. Ground water contamination caused by leachate infiltration from a dump.

Wells

Wells are used for a variety of purposes including exploration for mineral resources, drainage, disposal, and water supply. Eventually all are abandoned. Each of these activities may cause deterioration of water quality. Drainage wells are commonly used in swamps and potholes, allowing the excess water to flow into underground storage (Fig. 3-7). Sumps and dry wells may cause some local ground water pollution problems. They are typically installed to solve surface drainage problems, so they may transmit to ground water whatever pollutants are flushed into the well. For decades, human beings have disposed of liquid wastes by pumping them into wells. The injection of highly toxic wastes into some industrial deep-disposal systems has led to several water-pollution problems. Fresh water may be polluted by direct injection of wastes into the aquifer, by leakage of pollutants from the well head, through the casing, or via fractures in confining beds.

Another potential for pollution occurs in areas where abandoned and unplugged wells connect the disposal zone with shallow aquifers. Pressure injection may cause a considerable rise of the potentiometric surface.

A major cause of ground water pollution is the migration of mineralized fluids through abandoned wells. In many cases when a well is abandoned the casing (if there is one) is pulled or may become so corroded that holes develop. This permits ready access for fluids under high pressure to migrate either upward or downward through the abandoned well and pollute adjacent aquifers. Improperly cased wells allow high-pressure artesian saline water to spread from an uncased or partly cased hole into shallower, lower-pressure aquifers or aquifer zones, resulting in widespread salt intrusion.

Literally hundreds of thousands of abandoned exploratory wells dot the countryside. Many of these holes were drilled to determine the presence of underground mineral resources (seismic shot holes, coal, salt, oil, gas, etc.). The open holes permit water to migrate freely from one aquifer to another. A freshwater aquifer could thus be joined with a polluted aquifer or a deeper saline aquifer, or polluted surface water could drain into freshwater zones.

Improperly constructed water-supply wells may either pollute an aquifer or produce polluted water. Dug wells, which generally have a large diameter, shallow depth, and poor protection, are commonly polluted by surface runoff flowing into the well. Other problems have been caused by infiltration of water through polluted fill around a well, and still others by barnyard, feedlot, septic tank or cesspool effluent draining directly into the well. A wide variety of pollution and health problems can arise because of poor well construction.

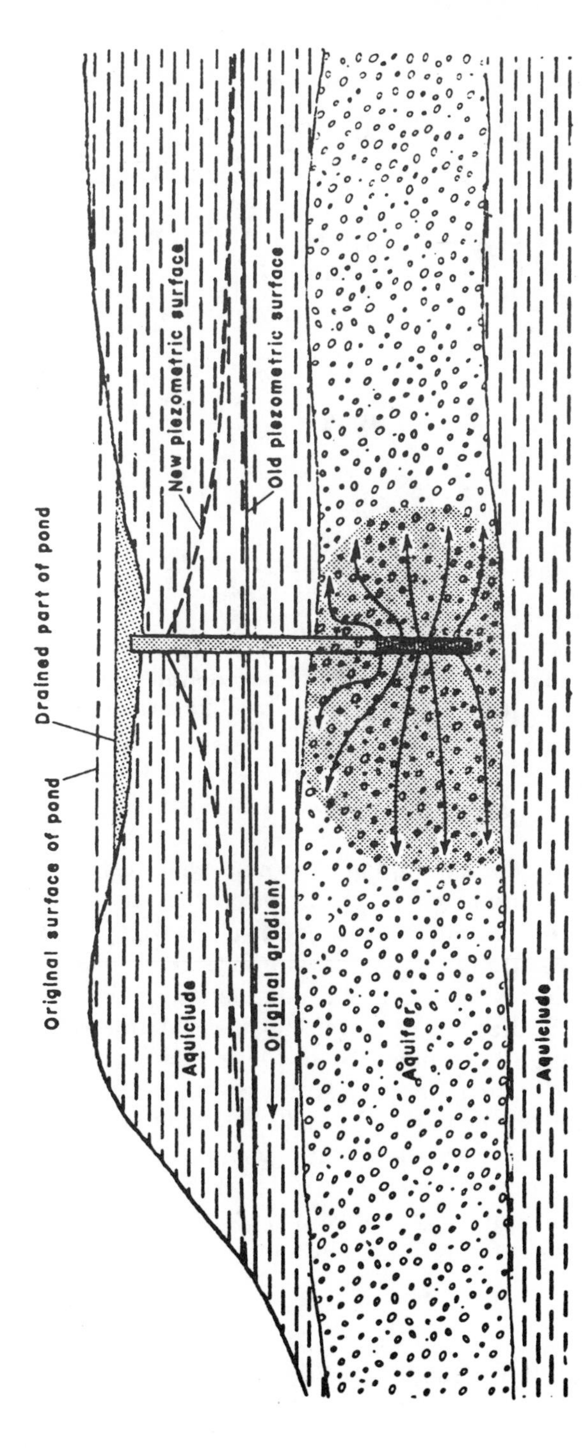

Figure 3-7. Drainage of a pond into an aquifer through a drain well. (*From Deutsch, 1965.*)

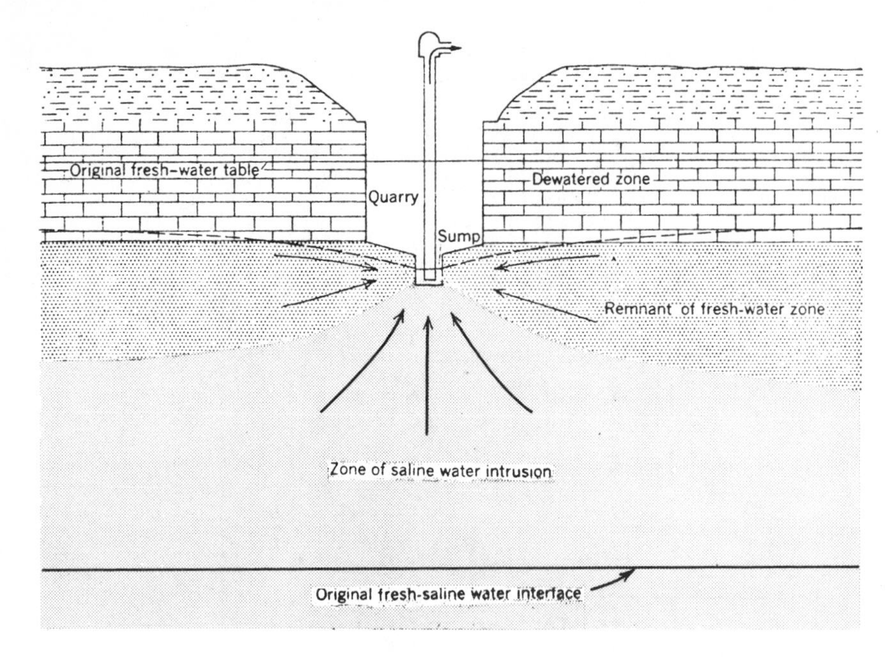

Figure 3-8. Migration of saline water caused by dewatering in a freshwater aquifer overlying a saline water aquifer. (*From Deutsch, 1963.*)

Ground Water Development

In certain situations, the pumping of ground water can induce significant water-quality problems. The principal causes include interaquifer leakage, induced infiltration, and landward migration of seawater (salt-water intrusion) in coastal areas. In these situations, the lowering of the hydrostatic head in the freshwater aquifer leads to migration of more highly mineralized water toward the well. In undeveloped coastal aquifers, the hydraulic gradient slopes toward the sea. Freshwater discharged from the aquifers through springs seeps into the ocean. Extensive pumping lowers the freshwater potentiometric surface and permits seawater to migrate toward the pumping center. A similar predicament occurs in inland areas where saline water is induced to flow upward, downward, or laterally into a freshwater aquifer because of the decreased head (pressure) in the vicinity of a pumping well (Fig. 3-8).

REFERENCES

Deutsch, Morris, 1963: Ground Water Contamination and Legal Controls in Michigan, U.S. Geological Survey Water-Supply Paper 1691.

Deutsch, Morris, 1965: Natural Controls Involved in Shallow Aquifer Contamination, *Ground Water*, vol. 3, no. 3, pp. 37–40.

Biesecker, J. E., and J. R. George, 1966: Stream Quality in Appalachia as Related to Coal-Mine Drainage, 1965, U.S. Geological Survey Circular 526.

Pettyjohn, W. A., 1972: Good Coffee Water Needs Body, *Ground Water*, vol. 10, no. 5, pp. 47–49.

Pettyjohn, W. A., 1975: Chloride Contamination in Alum Creek, Central Ohio, *Ground Water*, vol. 13, no. 4, pp. 332–339.

Miller, R. H., and W. A. Pettyjohn, 1974: Liquid Waste Disposal Areas *in* Great Lakes Pollution from Land Use Activities, Great Lakes Basin Comm., vol. 2, pp. 1–72.

Pettyjohn, W. A., J. R. J. Studlick, R. C. Bain, and J. H. Lehr, 1979: *Atlas of Drinking Water Quality from Ground Water Sources in Rural America*. National Water Well Assn., Worthington, Ohio.

4

Domestic Water-Supply-System Construction

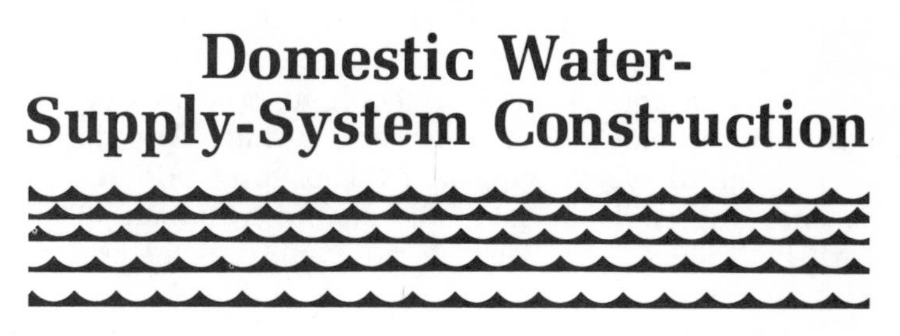

Adequate quantity, quality, and acceptable cost are the three major considerations when choosing a domestic water-supply source. For a modest fee, the homeowner can hire an engineer or hydrologist to evaluate the feasibility of developing a water source. Water-well contractors can assist by providing an estimate of the cost of ground water development. The U.S. Geological Survey, Soil Conservation Service, university extension services, and state and local health departments can also help the homeowner by providing information on resource reliability and system design and construction.

GROUND WATER SYSTEMS

Most rural and suburban homes are supplied by ground water from wells, springs, or hillside seepage. Other homes may use surface water from lakes, streams, and cisterns, but in arid and semiarid regions these sources often prove inadequate in both quantity and quality.

Ground water occurs in saturated strata called aquifers. These rock formations may be unconsolidated (sand or gravel), consolidated (sandstone, conglomerate, shale), or crystalline (granite, limestone, marble). Ground water is present nearly everywhere.

The water table lies at the top of a saturated aquifer, anywhere from a few to several hundred feet below land surface. The water table in all aquifers rises and falls a few feet annually. In some cases it may rise as much as 10 ft (3 m), causing wells to flow. These artesian aquifers occur where water is confined by an overlying bed of low permeability.

WELLS

Water wells are classified according to their method of construction: dug, bored, driven, jetted, and drilled. A dug well is simply dug by hand, with a bottom gravel covering and walls lined with bricks, stones, or culvert.

Dug wells are especially difficult to protect from contamination. They may become polluted by surface-water flow through an inadequately sealed well cover, or by seepage of polluted ground water. Because dug wells are generally shallow, they are also likely to fail during droughts, or when large quantities of water are rapidly withdrawn.

These wells are generally dug by hand with pick and shovel. A crib, culvert, or casing is used to prevent their caving in during excavation. When the well is completed, a few inches of gravel are spread over the bottom.

Bored wells are similar to dug wells but are generally deeper. Usu-

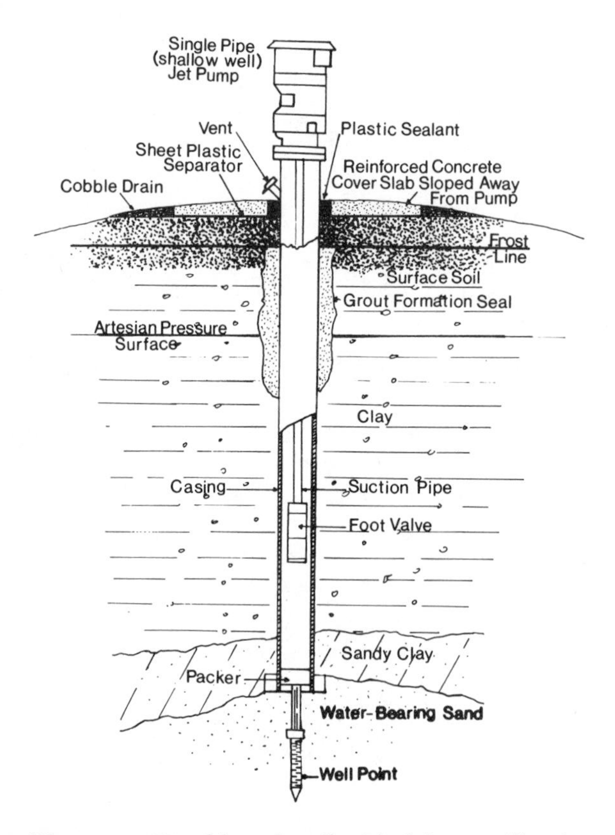

Figure 4-1. Hand-bored well with driven-well point and "shallow well" jet pump. (*U.S. EPA, 1973.*)

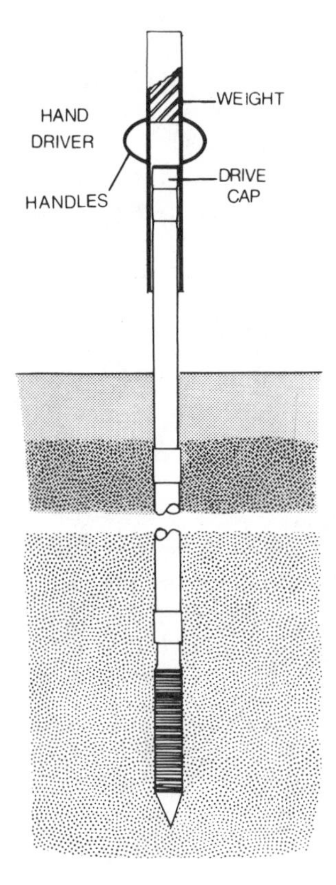

Figure 4-2. Simple tool for driving well pumps to depths of 15 to 30 ft (5 to 10 m).

ally lined with tiles, concrete pipe, wrought-iron or steel casing, these wells can be deepened if necessary by driving a well point through the bottom (Fig. 4-1). Proper protection of bored wells includes sealing the casing with cement grout to protect the well from surface-water pollution.

Driven wells are the least expensive and simplest to construct. The well point, attached to a steel pipe, is usually forged or cast steel 1¼ to 2-in (3 to 5 cm) in diameter (Fig. 4-2). The well is simply driven into the ground with a special weight or maul. The yield of driven wells is generally less than 7 gal/min (26.50 L/min).

Driven, dug, or bored wells seldom exceed 50 ft (15.25 m) in depth. They are usually more susceptible to pollution and may not always provide a reliable water supply.

Jetting is a quick and effective way to construct wells in soft, loose material, requiring only a water source and a pressure pump. The drill tool consists of a small-diameter standard pipe with a bit or chisel attached to the bottom section. Water is forced down through the drill pipe by means of the pressure pump and out through holes in the bit. This water, being under pressure, carries the cuttings to the surface through the space between the casing and the drill pipe. The lifting and dropping action of the drill pipe chops up the material in the hole and loosens it so that it may be washed to the surface. This method uses a short, fast stroke and is very effective in soft ground, sand, and gravel, or other loose unconsolidated formations. This method is best suited for smaller holes of from 2- to 4-in (5- to 10-cm) diameter. A properly constructed jetted well can be an excellent water source.

Drilled wells, which may exceed 1000 ft (300 m) in depth, are constructed with machine-operated drilling equipment or rigs. Since they usually provide the safest and most reliable water supply, they are generally preferred.

Two techniques used for drilled wells are percussion and hydraulic rotary drilling. In the percussion (cable-tool) method a heavy drill bit and stem is alternately raised and lowered to dislodge pieces of the formation. Water is mixed with the cuttings to form a slurry which is periodically bailed to the surface. Casing slightly larger than the drill bit is driven into the ground to prevent caving.

The rotary drilling method is equally common. Rigs include a derrick and hoist, a revolving table for drill pipe passage, a series of drill pipe sections, a cutting bit, a pump to circulate drilling fluid, and a power source to drive the drill.

As the rig drills, a bit rotates and advances to break up rock material. Drilling fluid (water or a special drilling mud) is pumped down the pipe and returns to the surface carrying cuttings which are charged into a settling pit. When the well is completed, the drill stem is withdrawn and the casing and screen are set. Casing then is installed and sometimes grouted in place (Fig. 4-3).

Other rotary drilling methods are air rotary and pneumatic hammer. The former is similar to the hydraulic method, with compressed air being used as drilling fluid. It allows rapid penetration in consolidated (hard-rock) formations but is poorly suited to drilling unconsolidated materials such as sand and gravel. Down-hole air hammers, (a combination of percussion and rotary drilling) are well suited to very hard formations.

Well Construction. When planning domestic well construction, the first thing to do is check state and local codes, because some potential domestic-supply sources may be prohibited. Other considerations for well-site selection include aquifer type, depth to the water-bearing zone, and convenience. It is also important that the well be free from flooding or pollution. Consequently, it should be located to minimize

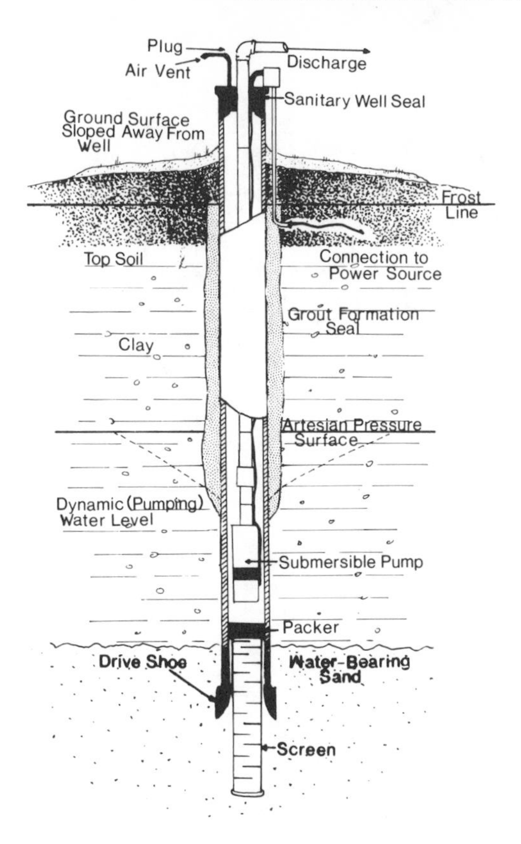

Figure 4-3. Drilled well with submersible pump. ***(U.S. EPA, 1973.)***

the threat of surface drainage to ponding and encourage the migration of potential ground water pollutants away from the proposed water source. The well should be accessible for maintenance, inspection, and repair. Its top should not be buried or located beneath a building, and the site should be well drained.

Casing. Casing should be surrounded by a dirt mound, extend above the ground level, and be new and strong enough to withstand installation force. Most well casing is steel pipe, although thermoplastic has proven satisfactory in some cases. Joints must be watertight—either welded or threaded. All casing should be covered with a sanitary well seal, cap, or pump mounting. The well should also be vented (vent pipe downward) and covered with a fine mesh screen (see Fig. 4-4).

Annular space between the hole and casing should be grouted with watertight cement or filled with clay to the necessary depth. A grout

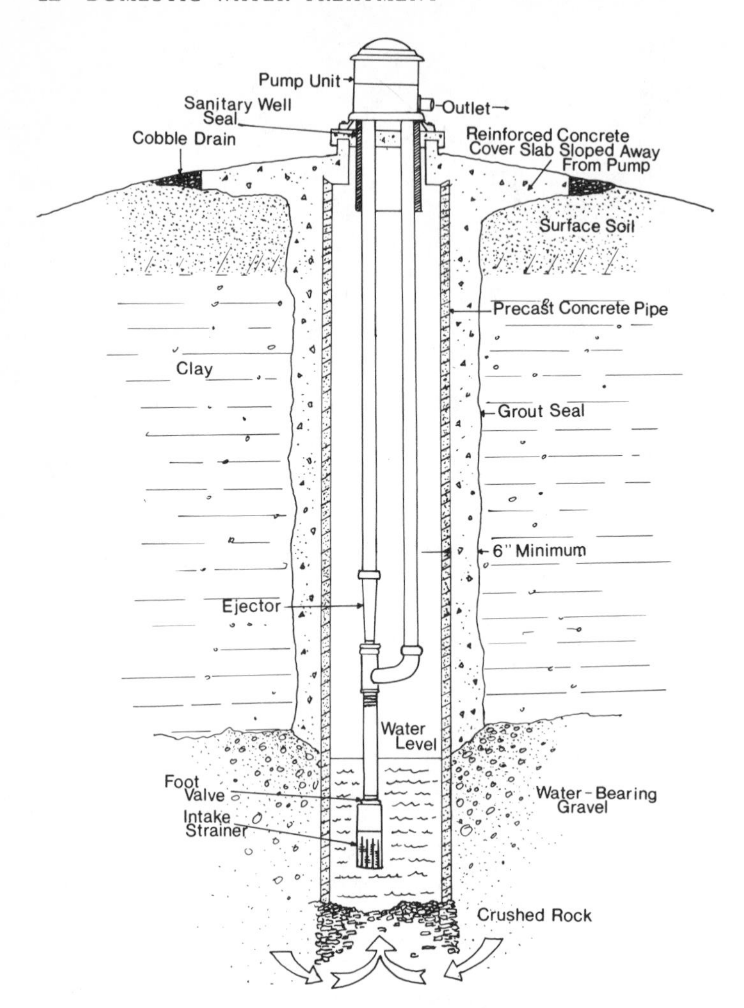

Figure 4-4. Dug well with two-pipe jet-pump installation. (*EPA, 1974.*)

seal is used to keep out polluted or highly mineralized water above the aquifer. When cementing is required, an oversized hole is drilled before the permanent well casing is installed. Filling the annular space between the casing and the wall of the bore hole is usually the last operation after the well has been test-pumped.

Well Development. Once construction is completed, the well should be "developed." This includes pumping the well and allowing the water to flow to waste. During this process, a contractor surges the well by turning the pump on and off every few minutes until the water is free of sand and turbidity. The repeated change in water-flow direction forces water and fine sediment into the well, where it is removed with the discharge water.

Removal of fine materials surrounding the well screen forms a size-graded envelope more permeable than the peripheral aquifer materials. Besides restricting infiltration of silt, clay, and sand, the envelope permits greater well yield with less drawdown.

Well Disinfection. Disinfection, the final step in well completion, kills any harmful organisms introduced during construction.

First, a concentrated chlorine solution is poured into the well, distributed throughout the entire water depth, and allowed to stand overnight. The amount of "dry" chlorine or bleach required for disinfection varies according to well diameter (Table 4-1). After disinfection, the well is pumped to flush out excess chlorine and then tested for the presence of bacteria. Wells should be disinfected each time they are serviced.

Well Maintenance

An owner should periodically check the well's water level and monitor any decline. Occasional preventive-maintenance inspections help to save money by allowing the owner to detect problems before they become serious.

Continued sanitary protection is also essential. Maintenance of the well seal and connections and protection from surface drainage are extremely important, since deterioration of well-safety components may allow it to become polluted.

SPRINGS AND SEEPS

A spring is a point where ground water flows at the surface. A seep is similar to a spring, except that water oozes rather than flows from the earth. Springs and seeps occur where water-bearing fractures, tunnels, or permeable zones obtrude from hillsides. Since their discharge varies, springs may become seeps or cease to flow during dry weather.

If properly protected, springs and seeps of sufficient discharge may provide a reliable water supply. Even springs which may not provide enough water for peak demand can fill a cistern in 24 h. Also, a supplemental pump can help provide for peak demand.

Two methods to secure and protect spring- or seep-water supplies are shown in Fig. 4-5. A ditch dug around the spring will divert surface drainage and help prevent pollution. A fence surrounding the site will

Table 4-1. Quantities* of Calcium Hypochlorite, 70% (rows A) and Liquid Household Bleach, 5.25% (rows B) Required for Water-Well Disinfection

Depth of water in well ft		*Well diameter, in.*															
		2	*3*	*4*	*5*	*6*	*8*	*10*	*12*	*16*	*20*	*24*	*28*	*32*	*36*	*42*	*48*
5	A	1T	1T	1T	1T	1T	1T	2T	3T	5T	6T	3 oz	4 oz	5 oz	7 oz	9 oz	12 oz
	B	1C	1C	1C	1C	1C	1C	1C	1C	2C	4C	1Q	2Q	3Q	3Q	4Q	5Q
10	A	1T	1T	1T	1T	1T	2T	3T	5T	8T	4 oz	6 oz	8 oz	10 oz	13 oz	1½ oz	1½ lb
	B	1C	1C	1C	1C	1C	1C	2C	2C	1Q	2Q	3Q	4Q	4Q	6Q	8Q	2½Q
15	A	1T	1T	1T	1T	2T	3T	5T	8T	4 oz	6 oz	9 oz	12 oz	1 lb	1½ lb	1½ lb	2 lb
	B	1C	1C	1C	1C	1C	2C	3C	4C	2Q	2½Q	4Q	5Q	6Q	2G	3G	4G
20	A	1T	1T	1T	2T	3T	4T	6T	3 oz	5 oz	8 oz						
	B	1C	1C	1C	1C	1C	2C	4C	1Q	2½Q	3½Q						
30	A	1T	1T	2T	3T	4T	6T	3 oz	4 oz	8 oz	12 oz						
	B	1C	1C	1C	1C	2C	4C	1½Q	2Q	4Q	5Q						
40	A	1T	1T	2T	4T	6T	8T	4 oz	6 oz	10 oz	1 lb						
	B	1C	1C	1C	2C	2C	1Q	2Q	2½Q	4½Q	7Q						
60	A	1T	2T	3T	5T	8T	4 oz	6 oz	9 oz								
	B	1C	1C	2C	3C	4C	2Q	3Q	4Q								
80	A	1T	3T	4T	7T	9T	5 oz	8 oz	12 oz								
	B	1C	1C	2C	4C	1Q	2Q	3½Q	5Q								
100	A	2T	3T	5T	8T	4 oz	7 oz	10 oz	1 lb								
	B	1C	2C	3C	1Q	1½Q	2½Q	4Q	6Q								
150	A	3T	5T	8T	4 oz	6 oz	10 oz	1 lb	1½ lb								
	B	2C	2C	4C	2Q	2½Q	4Q	6Q	2½G								

* Quantities are indicated as: T = tablespoons; oz = ounces (by weight); C = cups; lb. = pounds; Q = quarts; G = gallons. NOTE: Figures corresponding to rows A are amounts of solid calcium hypochlorite required; those corresponding to rows B are amounts of liquid household bleach. For cases lying in the shaded area, add 5 gal chlorinated water, as final step, to force solution into formation. For those in white area, add 10 gal chlorinated water.

Source: U.S. EPA, 1973

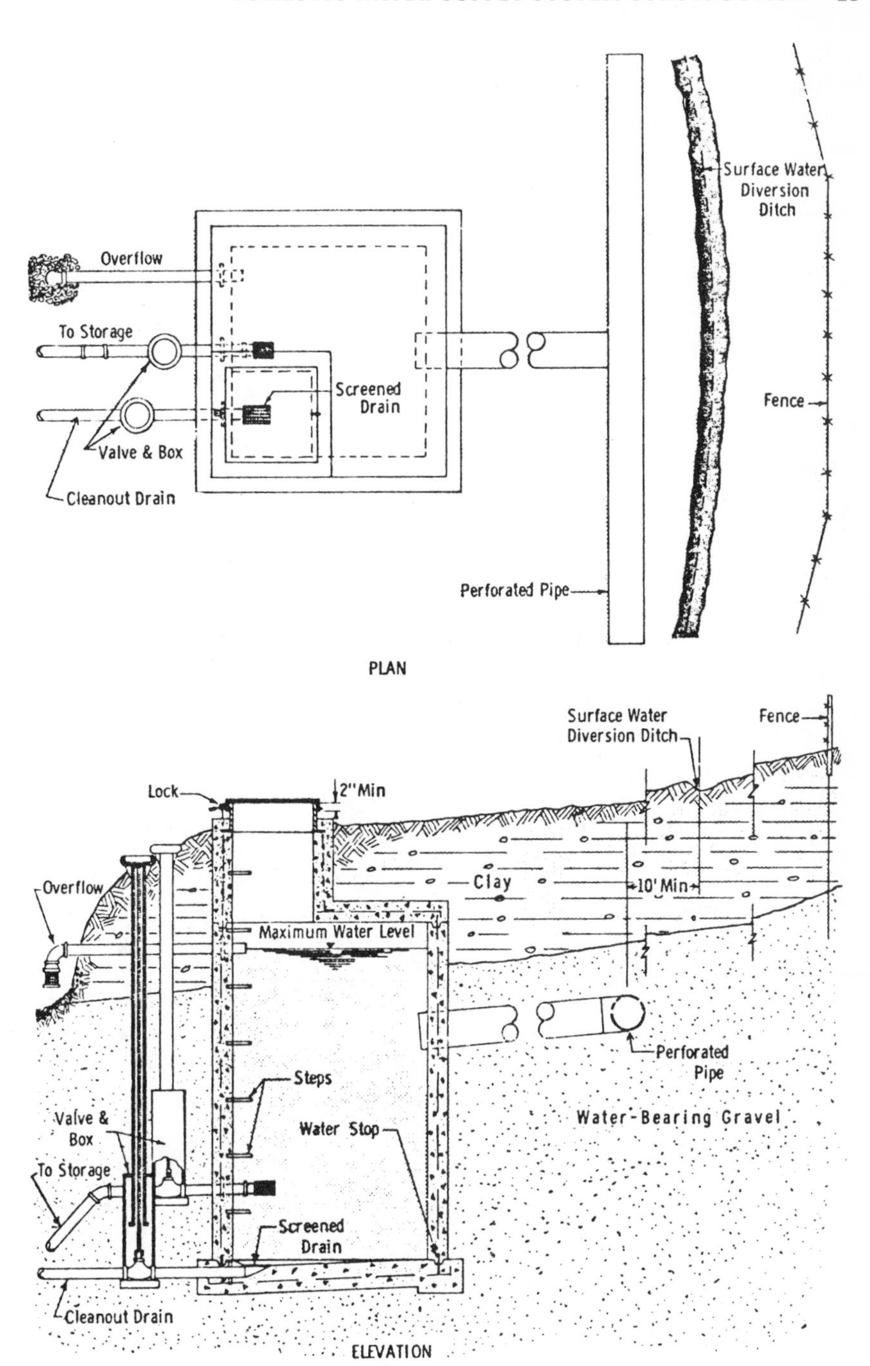

Figure 4-5. Spring protection.

bar entry by livestock. A reservoir covering will also help prevent such problems. After construction of the reservoir, the water supply should be disinfected. Furthermore, spring water should be checked periodically for contamination.

Springs can be contaminated by barnyards, sewers, septic tanks, or cesspools, especially when livestock are permitted free access to them. Some formations, such as limestone, facilitate the transmission of pollutants to springs. Increased turbidity following rainfall indicates that there is insufficient filtration. Therefore, it is likely the spring is contaminated.

SURFACE WATER

Cisterns

A cistern is a reservoir for collecting rainwater from roofs, or ground water from seepage zones, small springs, and low-yield wells. Available precipitation and roof runoff can be calculated with the aid of the chart in Fig. 4-6. Simply measure the length and width of the house and multiply the values to derive the roof's area in square feet.

A cistern should be located away from surface drainage and at least 50 ft (15.25 m) from sewers, septic tanks, cesspools, vault privies, stables, manure piles, etc. It should be vented, and its inlet and outlet pipes should be tightly sealed.

Underground cisterns are preferred, since they help maintain water temperatures and offer protection from freezing. But if satisfactory

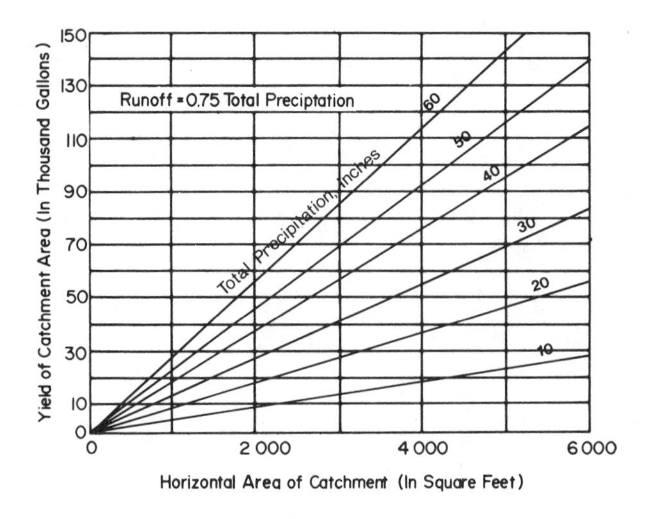

Figure 4-6.

water temperature can be maintained, they may be located above ground. This is usually the case where excavation is difficult, or where flooding is likely. To prevent settling and cracking, all cisterns should have a firm foundation.

Cistern-filtration methods include soft brick or cinder-block walls, sand, gravel, and charcoal filters for intake pipes, or floating filters with fiberglass inserts.

Since roof runoff is the major water source for cisterns, several possibilities for pollution exist. Rough surfaces will collect dirt and debris. Some roofs have soluble toxic coatings which can flow from the roof into the cistern. The entire gutter-downspout system should consist of materials approved for potable water systems use and should be securely anchored. The connecting downspouts should be able to withstand heavy rains. Gutters should be covered with screen or mesh. Leaks which could permit unfiltered water to mix with the filtered supply should be prevented. In addition, the first flush of rain should be diverted from the cistern through an automatic flapper valve on the inlet pipe.

Following its installation, the cistern must be disinfected. Elaborate precautions observed in cistern construction, maintenance, and repair still do not eliminate pollution potential. Automatic chlorination of water leaving the cistern is essential for a safe supply.

Figure 4-7 shows a properly constructed cistern. The manhole is correctly placed for visual inspection of the entire interior. It extends at least 3 in (7.5 cm) above the surface. The manhole cover is securely hinged and locked.

Streams, Lakes and Ponds

Careful consideration of drainage-basin conditions and precipitation must precede construction of surface-water storage facilities. Here are some questions that must be considered when one is planning a surface-water facility:

- Is the water quality acceptable?
- Are there contamination sources upstream?
- How much water will be lost through evaporation, transpiration, and seepage?

Catchment areas larger than 2 to 3 acres should be designed by competent engineers. The U.S. Weather Service, local soil conservation and agricultural agencies, and health and resources departments can help in the design of a suitable system.

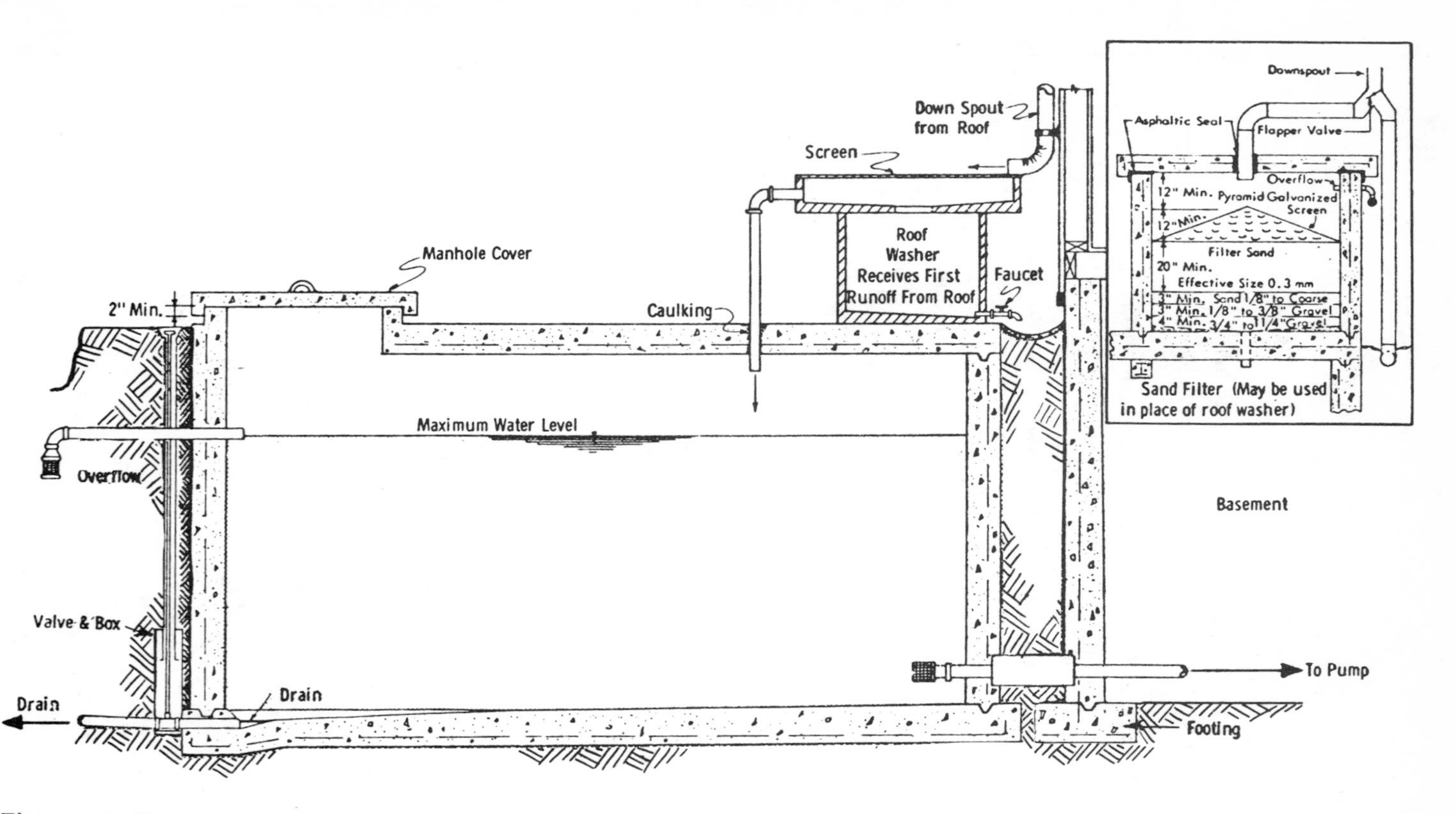

Figure 4-7. Cistern.

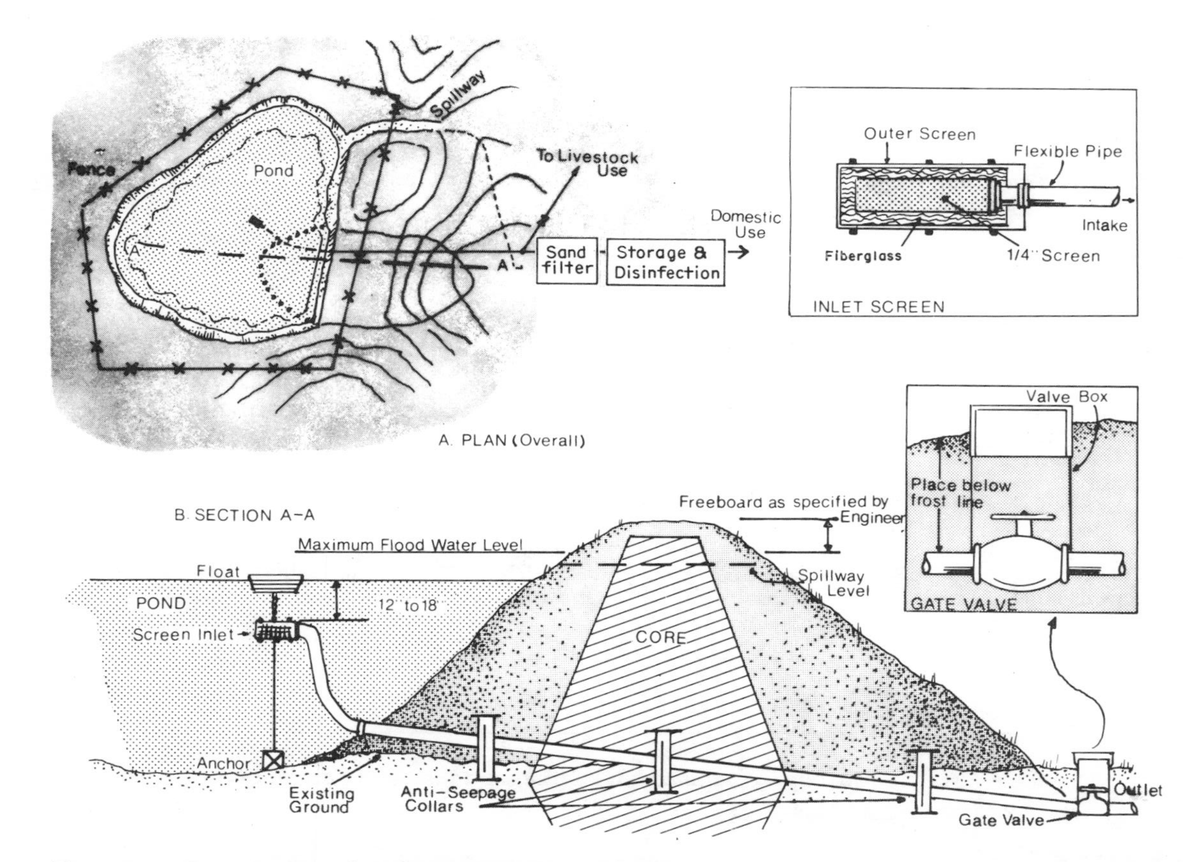

Figure 4-8. Construction of surface-water storage facility.

Table 4-2. Characteristics of Individual Water Supplies and Recommended Treatment

Source	*Adequacy of supply*	*Potability of water**	*Treatment recommended for domestic use*
Wells			
Drilled	Usually adequate; depends on geological formation and skill of driller	Nearly always potable when constructed and located to prevent pollution	Rarely requires treatment with proper construction; exception: water-bearing formations which have already been polluted should be continuously disinfected
Bored	Same as above	More susceptible to pollution by surface water when casing joints are unsealed; otherwise, usually potable	If bacteriological examination shows contamination, continuous disinfection of water is recommended
Jetted	Adequacy depends on nature of water-bearing formations, skill of driller, design of inlet structure, and final development (cleanup) of well-inlet structure	Nearly always potable; nature of formations in which jetting is possible usually affords good protection against bacterial contamination	Treatment rarely required; completion in very shallow formation where sewage disposal facilities, stockyards (feedlots), or commercial fertilizers are present can produce nitrate pollution; treatment of water to remove nitrates advisable only for small quantities of water needed for infant formula
Driven	Adequacy highly dependent on nature of water-bearing formations, design of drive (sand) point, depth below water table, and final development (cleanup) of inlet screen	Same as above (jetted)	Same as above (jetted); drive points sunk into shallow formations composed of coarse materials (sand and gravel) more susceptible to bacterial contamination; if bacteriological examination shows contamination, continuous disinfection of water is recommended

Dug	Generally shallow; low yield or dry in summer	Subject to pollution, particularly if poorly designed, improperly constructed, or inadequately protected: dug wells are not recommended for use as a source of drinking water	Automatic chlorination, frequent inspection of well, and periodic water-quality analysis recommended
Springs	Highly dependent on nature of formation supplying the spring; may be undependable if spring does not flow or if its flow fluctuates seasonally	Springs issuing from sandy formation and from deep rock formations more likely to be bacteriologically safe; those found in limestone and basalt not covered with thick overburdens of fine-grained materials are more likely to be contaminated; any spring that becomes turbid or whose flow increases noticeably following rainstorms is very likely to be contaminated	If spring is only source of water (because a well cannot be constructed) and bacteriological examination shows water to be contaminated, continuous disinfection is recommended
Cisterns	Rarely adequate to satisfy all demands imposed on it by modern home requirements; adequacy dependent on size of rain catchment area (usually roofs), storage capacity of cistern, and amount and distribution of rainfall	Likely to be contaminated with bacteria from bird and rodent droppings	If rainwater is only utilizable source of water (because a well cannot be constructed), continuous disinfection of the water is recommended
Streams, lakes, and ponds	Adequate except for intermittent streams	All surface sources likely to be polluted to some degree; high probability that water will be turbid after rains	Filtration and disinfection are necessary in most cases; if industrial or agricultural wastes are present, additional treatment may be necessary; consider possible presence of nitrates if streams or ponds receive drainage from fertilized fields

* Most potability problems in individual home water supplies result from the presence of microorganisms (bacteria, viruses, etc.). Rarely do toxic minerals appear in these supplies. Most mineral problems in ground water are merely nuisances and are not dangerous to health: for example, excessive iron concentrations, leading to "red water," and staining of clothes and fixtures. It is always advisable, however, to have a water sample analyzed by a laboratory to determine the possible presence of substances in concentrations exceeding those recommended by the state health department. (See Chap. 6.)

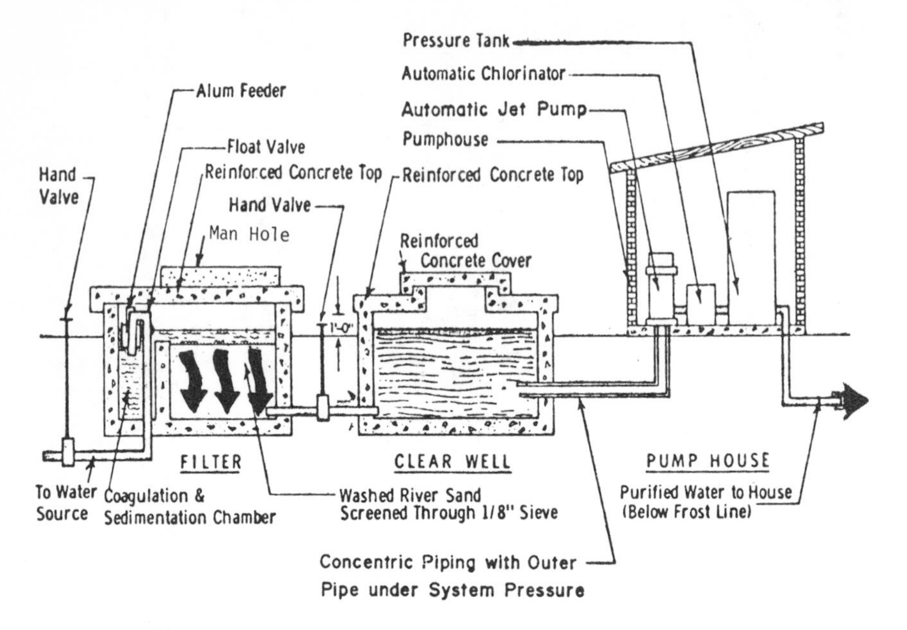

Figure 4-9. Pond-water-treatment system.

The most convenient and least expensive surface-water sources are streams and lakes, but ponds can also be constructed. Intake for pond, stream, or lake withdrawal is shown in Fig. 4-8.

Surface water can be polluted by various sources and is not considered safe without treatment. In addition to treatment for excess mineral content and acidity, surface water may require sediment, odor, and taste control. Drinking water from ponds requires more extensive treatment than water from any other source. One method of treating surface-water supplies is shown in Fig. 4-9. Table 4-2 lists recommended treatment of surface-water sources and various types of wells.

Ponds

A pond is designed to store water for use during periods of low precipitation and should contain a minimum of a year's estimated water supply, including allowances for seepage and evaporation. It should have a maximum depth of at least 8 ft (2.5 m), with an average water depth of at least 3 ft (1 m). Graphs similar to the one shown in Fig. 4-6 are available to approximate necessary catchment area.

Fences will protect a pond and help keep it free from weeds and debris. The pond's watershed must also be protected: it should be located in a grassy area away from barns and septic tanks. A watershed should never be treated with herbicides, pesticides, or fertilizers.

Further conservation measures may be necessary to maintain adequate water supplies in arid regions. Reducing catchment-area infiltration, plant water loss, pond seepage, and evaporation will all help increase water supply. Infiltration can be reduced by altering the surface of the catchment area and by chemical treatment of the soil. Pond seepage can also be reduced through chemical treatment. Water loss through plant transpiration can be reduced by removing plants which consume large quantities of water, such as willow, salt cedar, cottonwood, alder, mesquite, and box elder. Evaporation can be minimized by covering the water surface with liquid chemicals or blocks, rafts, or beads. Water storage in sand and rock dams also helps minimize evaporation loss.

REFERENCES

American Association for Vocational Instructional Materials, 1973: *Planning for an Individual Water System.* Athens, Ga.

Bailey, N. G., 1959: *Cisterns for Rural Water Supply in Ohio,* Ohio Department of Natural Resources.

Campbell, M. D., and Jay H. Lehr, 1973: *Rural Water Systems Planning and Engineering Guide,* Commission on Rural Water. Washington, D. C.

E. E. Johnson, 1966: *Ground Water and Wells.* Johnson, St. Paul, Minn.

U.S. Environmental Protection Agency (EPA), 1973: *Manual of Individual Water Supply Systems,* U.S. EPA, Office of Water Programs, Water Supply Division, EPA-430/9-74-007.

5

Significance of Water-Quality Constituents

An examination of water analyses clearly shows that the major dissolved substances in water comprise a small number of elements, usually in ionic form. These include the ions of calcium, magnesium, sodium, bicarbonate, sulfate, and chloride. Although it rarely happens in a water sample, these substances can combine to form a total of nine dissolved chemical compounds. There may be several other compounds in water, but they are generally in much smaller quantities. Mineral substances found in low concentrations include iron, manganese, silica, nitrate, and fluoride as well as scores of others, including a wide variety of trace elements. Unfortunately, even minor concentrations of some of these substances can produce undesirable taste, color, odor, or staining, or can even adversely affect health. Because of the undesirable characteristics of some of the substances commonly found in water, the U.S. Environmental Protection Agency has established Primary Drinking Regulations that limit the concentrations of substances permitted in public drinking water supplies. Although they were developed for public supplies, these regulations, promulgated as required by the Federal Safe Drinking Water Act of 1974, (Appendix—Drinking Water Standards) are useful for assessing the safety and general acceptability of water from individual home systems as well.

DISSOLVED SOLIDS

Dissolved solids and total dissolved solids are terms, commonly used interchangeably, that denote the concentration of mineral constituents dissolved in water. Dissolved solids do not include gases, colloids, or sediment, but consist chiefly of carbonates, bicarbonates, chlorides, sulfates, phosphates, and nitrates of calcium, magnesium, sodium, and potassium, with traces of iron, manganese and a few others (McKee and Wolfe, 1963, p. 181). Total solids include both suspended and dissolved solids, so the term is not interchangeable with dissolved solids.

The concentration of dissolved solids is determined or estimated by the specific conductance, the residue on evaporation of the filtrate, the sum of the concentrations of all the constituents determined by analysis, or less commonly, by specific gravity determinations. One common method of determining dissolved solids is to evaporate a known volume of water and weigh the residue. During the evaporation process, bicarbonate is changed to carbonate, carbon dioxide, and water, while some magnesium, chloride, nitrate, and organic materials are partly volatilized.

The computed dissolved-solids content is also equal to the sum of the major ions determined in each sample analysis, after all solid constituents are converted mathematically into the forms in which they would normally exist in a dry state. Obviously, partial chemical analysis may make it impossible to estimate dissolved solids because of the limited number of constituents measured.

Rarely will the dissolved solids content determined by evaporation be equal to that determined by calculation. Hem (1963, p. 220) suggests that the values may differ by ± 10 to 20 mg/L, when the concentration is on the order of 100 to 500 mg/L. Three reasons for this discrepancy are (1) the chemical analysis may not be sufficiently complete, (2) during evaporation several substances, such as organic materials, may be volatilized, and (3) some water of hydration may remain in the residue.

In solutions in which the concentration of dissolved solids is greater than 1000 mg/L, the calculated value may be preferable to the residue on evaporation. In some analyses the method used to determine dissolved solids is not reported. Where the concentration is less than 1000 mg/L, it is reported to the nearest whole number, but only to three significant figures at higher concentrations.

Although the National Secondary Drinking Water Regulations (1977) proposed a limit of 500 mg/L of dissolved solids in drinking water, concentrations in excess of 1000 mg/L are commonly found in municipal supplies where less mineralized waters are not available. In fact, concentrations as great as about 5000 mg/L are presently found in municipal supplies (Office of Saline Water, 1969), and concentrations ranging between 2000 and 4000 mg/L are not uncommon. Although the concentration of dissolved solids is used in a variety of water-classification schemes, the usefulness of the supply must be based on the concentration of individual ions rather than the total concentration of all substances.

The recommended limit for dissolved solids is based mainly on taste thresholds and not physiological effects. Drinking waters that contain more than 500 mg/L of dissolved solids may have a taste. Palatability decreases with increasing concentrations. As pointed out by Rainwater and Thatcher (1960, p. 269),

> Water with several thousand parts per million of dissolved solids is generally not palatable, although those accustomed to highly mineralized water may complain that less concentrated water tastes flat. More often, a change in the source of drinking water may cause gastric disturbances rather than the concentration of dissolved solids in the water itself.

Though some animals can temporarily survive on water containing 18 000 mg/L of dissolved solids, livestock and domestic animals may begin to be distressed when the level reaches 3000 mg/L. Therefore, concentrations should be less than 2500 mg/L. Crops can also be affected by highly mineralized water, and that used for irrigation should contain less than 1500 mg/L and preferably no more than 700 mg/L of dissolved solids; some plants may survive at concentrations in excess of 20 000 mg/L. Highly mineralized waters also trouble industry by causing foaming and scaling in boilers, accelerating corrosion, and interfering with cleanliness or taste in finished products.

The U.S. Geological Survey established an arbitrary system for ranking saline waters relative to their dissolved-solids content (Table 5-1). Although not used in this classification, the terms semibrackish and brackish (1000 to 10 000 mg/L) are also common.

Dissolved solids in natural waters range from less than 10 mg/L for rain to more than 100 000 mg/L in brines (Table 5-2). In the humid areas of North America, uncontaminated surface and ground waters generally contain less than 500 mg/L, but concentrations increase with depth and drier climates (Fig. 5-1). In semiarid and arid regions, dissolved solids from both surface and underground sources are likely to exceed 1000 mg/L; and in large areas, as much as 2000 mg/L are common. Even

Table 5-1. Classification of Saline Water

Classification	*Dissolved solids, mg/L*
Slightly saline	1000–3000
Moderately saline	3000–10 000
Very saline	10 000–35 000
Briny	More than 35 000

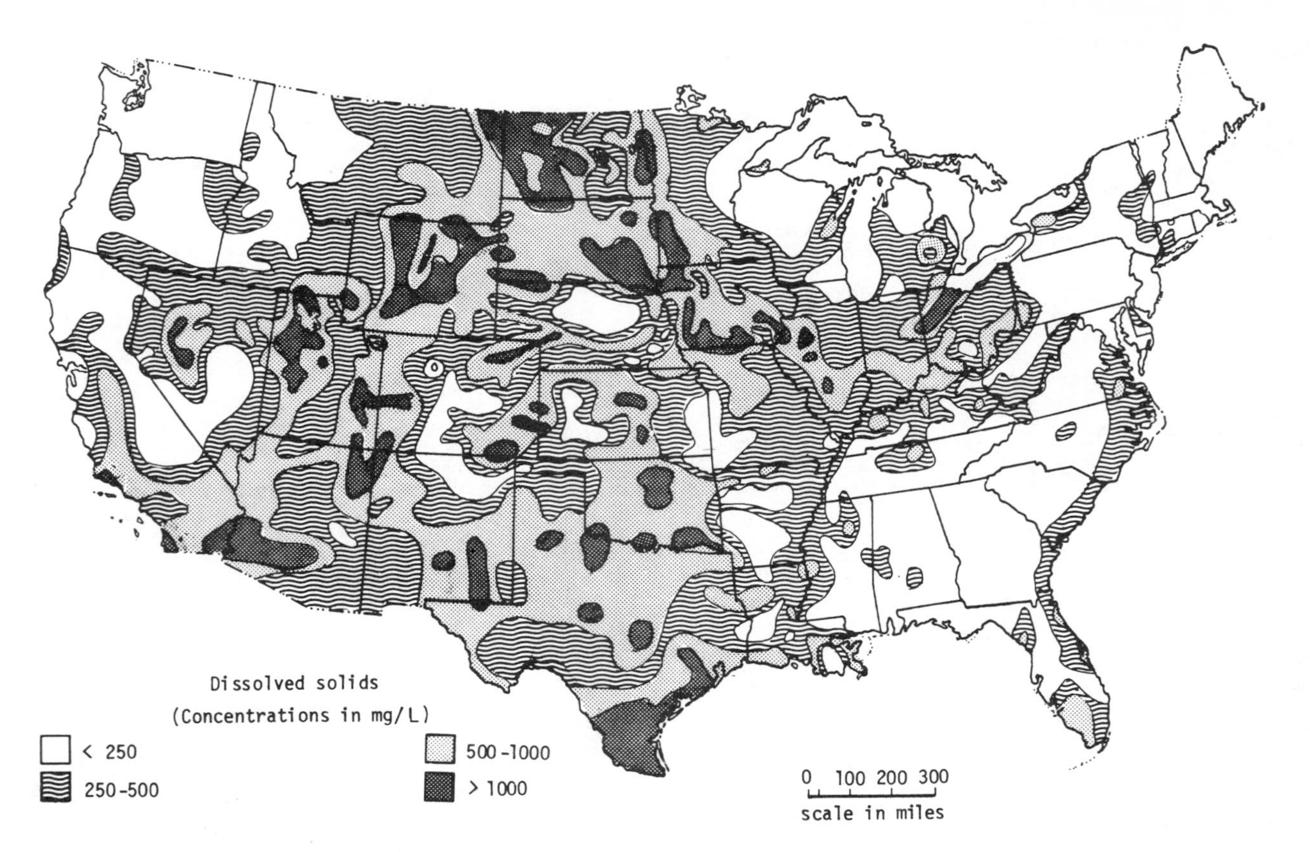

Figure 5-1. Dissolved solids in untreated ground water used for drinking supplies.

Table 5-2. Range in Concentration of Dissolved Solids from Several Sources

Source	*Dissolved solids, mg/L*
Distilled	0
Rain and snow	10
Lake Michigan	170
Average of all rivers in United States	210
Missouri River	360
Pecos River	2600
Ocean	35 000
Brine well	125 000
Dead Sea	250 000

higher concentrations are encountered where evaporites lie at or near land surface.

The chemical quality and sediment load in rivers and streams, particularly in arid and semiarid regions, can fluctuate within wide extremes (Table 5-3). Generally, the lowest concentrations exist where

Table 5-3. Variations in Dissolved Solids, Chemical Type, and Sediment Rivers in Arid and Semiarid Areas in the U.S. (FWPCA, 1968, p. 168)

	Dissolved solids concentrations, mg/L			*Sediment concentrations, mg/L†*	
Region	*From*	*To*	*Prevalent chemical type**	*From*	*To*
Columbia River Basin	<100	300	Ca-Mg, C-b	<200	300
Northern California	<100	700	Ca-Mg, C-b	<200	+500
Southern California	<100	+2000	Ca-Mg, C-b; Ca-Mg, S-C	<200	+15 000
Colorado River Basin	<100	+2500	Ca-Mg, S-C; Ca-Mg, C-b	<200	+15 000
Rio Grande Basin	<100	+2000	Ca-Mg, C-b; Ca-Mg, S-C	+300	+50 000
Pecos River Basin	100	+3000	Ca-Mg, S-C	+300	+7000
Western Gulf of Mexico Basins	100	+3000	Ca-Mg, C-b; Ca-Mg, S-C; Na-P, S-C.	<200	+30 000
Red River Basin	<100	+2500	Ca-Mg, S-C; Na-P, S-C	+300	+25 000
Arkansas River Basin	100	+2000	Ca-Mg, S-C; Ca-Mg, C-b; Na-P, S-C.	+300	+30 000
Platte River	100	+1500	Ca-Mg, C-b; Ca-Mg, S-C	+300	+7000
Upper Missouri River Basin	100	+2000	Ca-Mg, S-C; Na-P, C-b; Na-P, C-b.	<200	+15 000

* Ca-Mg, C-b: Calcium-magnesium, carbonate-bicarbonate. Ca-Mg, S-C: Calcium-magnesium, sulfate-chloride. Na-P, C-b: Sodium-potassium, carbonate-bicarbonate. Na-P, S-C: Sodium-potassium, sulfate-chloride.

† $\text{Sediment concentration} = \frac{\text{annual load}}{\text{annual streamflow}}$

the surface water is diluted by snow melting and surface runoff derived from rains.

SPECIFIC ELECTRIC CONDUCTANCE

Specific conductance, electric conductance or conductivity, is an extremely useful measurement that is both conveniently and rapidly determined. This measurement, indicated by a meter, is used to estimate the concentration of dissolved solids in water. Conductance, the ability of a substance to conduct an electric current, is the reciprocal of the resistance of a cube of the substance 1 cm on a side at a specific temperature, usually 25°C. It is reported in units of siemens* (S), but since natural waters have conductivity values far less than 1 S, the data generally are reported in units of microsiemens (μS) or millionths of a siemens.

Dissolved solids can be estimated by multiplying the conductivity by some predetermined constant (C). Dissolved solids, mg/L = C $\times$ specific conductance, μS. If the water is highly mineralized, dissolved solids are usually more than 65 percent of the conductivity, but they are less if the water has a high or low pH or contains sodium chloride. As a general rule, the conductivity is multiplied by 0.65 in order to estimate the dissolved solids.

The reason for this difference is that there is no simple relation between ionic concentration and conductance. How well a current in a given solution will be conducted depends on the number and kinds of ions present, their relative charge, and the freedom of ions to act as conductors. Dissolved-solids concentrations should not be estimated from conductivity values that exceed 50 000 μS, because the relation becomes indefinite for solutions approaching saturation.

Distilled water generally has a conductance that ranges between 1 and 5 μS, while rainwater commonly ranges from 10 to about 50. Conductance may be considerably higher, however, where the atmosphere is polluted with sulfur dioxide or other industrial contaminants, in coastal areas where it may contain sea salt, or in arid regions where windblown dust from evaporite deposits is prevalent. Elsewhere, the conductivity of both surface and ground waters varies widely from nearly distilled water to brine. Of course, the conductivity of ground water at a particular site is nearly constant, while the conductance of surface water varies with the discharge of the stream.

SODIUM

The major source of sodium in natural waters is from the weathering of feldspars, evaporites, and clay. Many sodium salts are also used in

* Formerly called a mho.

industry. Sodium and its salts are very soluble and tend to remain in solution, in contrast to iron, calcium, and magnesium. Unlike calcium, there are no important precipitation processes that can maintain low sodium concentrations in water; since sodium salts are so soluble, they commonly will not precipitate unless concentrations of several thousand milligrams per liter are reached. Most sodium-rich deposits are found in desert basins where they have been concentrated by evaporation.

Sodium is retained by adsorption on surfaces of high cation-exchange–capacity minerals, such as clay, but it may be either added to or subtracted from water by ion exchange. This particular reaction is reversible: calcium can replace sodium or sodium can replace calcium, the direction of change depending on the concentration of the ions. In many natural systems, ground water, as it slowly migrates through fine-grained material, is softened as calcium ions are replaced by sodium ions.

Probably all natural waters contain some sodium, concentrations of which range from less than 0.5 mg/L in rain to more than 1000 mg/L in brines. Ground water in rocks of low solubility, such as most igneous and metamorphic rocks, will generally contain sodium concentrations that range between 1 and 20 mg/L. Similarly, concentrations are usually small in carbonate aquifers where calcium is a dominant ion, but they may increase to more than 100 mg/L in aquifers that have a high ion-exchange capacity, such as clay, sand, sandstone, and siltstone. Furthermore, the sodium content in some well waters may increase with time, due to leakage from adjacent clay-rich rocks.

Sodium is not especially important in water for domestic uses, but persons with an abnormal sodium metabolism should consult their physicians concerning the planning of a low-sodium diet if the supply has a high-sodium content. Since there are no apparent adverse physiological effects on healthy people caused by sodium, a recommended limit in drinking waters has not been established, but sodium in excess of 500 mg/L, when combined with chloride, produces a salty taste.

POTASSIUM

Although both elements have similar chemical properties, potassium is usually less abundant than is sodium in natural waters. Potassium reacts to form clays, while sodium tends to remain in solution. Common salts of potassium are very soluble and not easily removed from water by most processes. Concentrations, normally less than 20 mg/L in natural waters, may be as much as 100 mg/L in hot springs and exceed 25 000 mg/L in brines.

Potassium is an essential nutrient, but large concentrations are laxative. From the point of view of domestic water supply, potassium is of little importance and creates no adverse effects. There is presently no recommended limit in drinking-water supplies.

CALCIUM

Calcium in its elemental form does not occur in nature, but calcium ions are the principal cations in most natural waters. It is widely distributed in soil and rock and its presence in water may be derived from the leaching of soil, the solution of carbonate rocks, or the disposal of sewage and many types of industrial wastes. It is extremely mobile in the hydrosphere. Although calcium is derived from nearly all rocks, the greatest concentrations are derived from limestone, dolomite, gypsum, and gypsiferous shale.

Cation-exchange equilibria have considerable influence on calcium concentration in most natural waters. The exchange of calcium for sodium in an aquifer is known as natural softening. It is common in many areas, particularly where a large ion-exchange capacity is available, for example, in shale and clay. Although less common, the exchange of sodium for calcium also occurs.

Concentrations of calcium in natural water commonly range between 10 and 100 mg/L. Rocks of low solubility, such as granite and siliceous sandstone, may contain less than 10 mg/L, limestone waters may range from 30 to 100 mg/L, and the calcium content of water from gypsum-rich strata commonly exceeds 100 mg/L. Some calcium chloride brines, which in themselves are unusual, may exceed 50 000 mg/L of calcium.

Limits on calcium in water would be desirable for domestic supplies, not because it is a health hazard, but because it may be disadvantageous for many household uses, such as washing, bathing, and laundering. Small concentrations may be desirable, however, because they tend to inhibit corrosion of plumbing by the formation of a protective coating (scale) in pipes. Since calcium is the major cause of hardness, its combination with the fatty acids in soap forms a curd, which collects on sinks, tubs, and clothing. Furthermore, calcium and magnesium combine with bicarbonate, carbonate, sulfate, and silica to form a heat-retarding, insulating scale in boilers and other heat-exchange equipment.

Humans require 700 to 1200 mg/L of calcium per day, and some investigators believe that lack of calcium is the most common nutritional deficiency in the United States. Drinking water can provide only a small fraction of the required calcium, however. Most must come from foods. While several physiological effects of calcium have been suspected, they bear no definite relationship to calcium in drinking water. It does not cause hardening of the arteries, kidney stones, or liver ailments. Concentrations as great as 1800 mg/L of calcium water are reported to be harmless. In fact, calcium in water reduces the toxicity of many chemical compounds to fish and other aquatic organisms.

MAGNESIUM

Sources of magnesium include ferromagnesium minerals in igneous and metamorphic rocks and magnesium carbonate in limestone and dolomite. The geochemistry of magnesium is similar to that of calcium and its solubility is controlled largely by carbon dioxide. In natural waters, there is usually less magnesium than calcium. The ratio of calcium to magnesium in seawater is about 5 to 1; calcium is normally removed by plants and animals to build their supporting structures. In aquifer systems where natural softening has occurred, magnesium concentrations are also low because clay tends to exchange sodium for both calcium and magnesium.

Humans require about 350 mg/L of magnesium in their daily diets. It is relatively nontoxic to human beings and is not a health hazard. The Public Health Service Standards of 1946 recommended a maximum concentration of 125 mg/L of magnesium because magnesium salts can serve as cathartics and diuretics on new users. This limit was removed from later standards; with continued consumption, most individuals can develop a tolerance to the higher concentrations within a short time. Magnesium also produces an unpleasant taste, the threshold being about 500 mg/L for most individuals; water becomes unpalatable long before toxic concentrations are reached.

Magnesium concentrations in water from rocks of low solubility, such as granite and sandstone, are generally less than 5 mg/L, and range from 10 to 50 mg/L in water from limestone and dolomite. Amounts greater than 100 mg/L are rare. Magnesium may be enriched in cave waters where calcite has been precipitated.

HARDNESS

Hard water and soft water are terms that have no exact meaning because water considered hard in one region might be considered soft by inhabitants in another region. Hardness is usually associated with what happens when soap is used. Soap does not clean as efficiently in hard water. It leaves insoluble residues in bathtubs, sinks, and clothing. In addition, hard water causes scale to incrust water heaters, boilers, and pipes thereby reducing their capacity and heat-transfer properties. Thus, it is a technical and economic problem.

Hardness is commonly reported in units of grains per gallon (gpg) or as milligrams per liter as $CaCO_3$, although other terms are also used, as follows:

1 gpg (U.S.) = 17.12 mg/L
1 French degree = 10 mg/L
1 German degree = 17.8 mg/L
1 English or Clark degree = 14.3 mg/L

Grains per gallon, an ancient concept, was originally defined as the degree of hardness produced by a grain of chalk in a gallon of water.

For the most part, hardness depends on the concentration of calcium and magnesium. Hardness is usually expressed in terms of the equivalent quantity of calcium carbonate ($CaCO_3$). There are no distinctly defined levels for what constitutes a hard- or soft-water supply. However, a general scale of hardness (as $CaCO_3$) has been provided by the Water Quality Association as follows:

Concentration		
mg/L	*gpg*	*Description*
0–17	0–1	Soft
17–60	1–3.5	Slightly hard
60–120	3.5–7	Moderately hard
120–180	7–105.	Hard
over 180	over 10.5	Very hard

In some analyses, hardness is reported as carbonate, or temporary, hardness and noncarbonate, or permanent, hardness. Carbonate hardness is considered temporary because it can be reduced by boiling the water. Noncarbonate hardness is not reduced by boiling—hence the term permanent.

Water hardness ranges within wide limits (Fig. 5-2). Lake Superior contains about 68 mg/L, Lake Michigan contains about 128 mg/L, and Ohio streams range between 200 and 500 mg/L. Although the chemical quality of ground water is relatively constant, the hardness in streams varies seasonally. The hardness in the Mississippi River at Moline, Illinois, for example, doubles in the wintertime, from 137 to 274 mg/L. One of the major reasons for this variability of hardness in a stream is the changeable percentage of ground water runoff, which is usually more mineralized than is surface runoff.

Hardness in water does not present a health hazard and may, in fact, be advantageous in some cases. For example, in soft, corrosive water areas where lead pipes are used, a scale does not form. Naturally, soft, corrosive waters may dissolve the lead and continual consumption of the water may present a serious health hazard. Even in newly plumbed

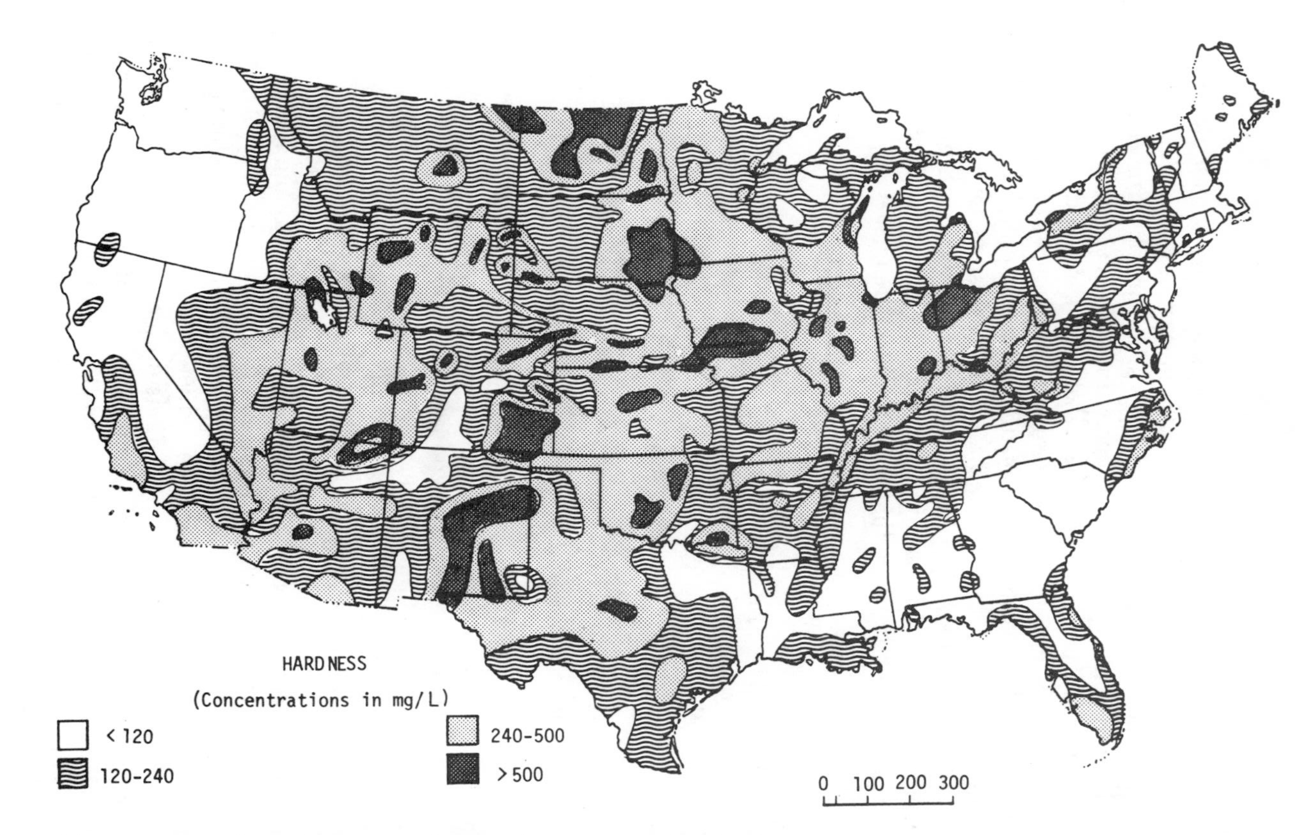

Figure 5-2. Hardness in untreated ground water used for drinking supplies.

homes in both soft- and hard-water areas, the concentration of copper in water may be unusually high until the scale has an opportunity to form and protect the bare metal.

Rocks and soils that contain large concentrations of calcium and magnesium can contribute large concentrations of hardness. Hard water can be expected in terrains underlain by carbonate rocks, in glaciated areas where the till contains large concentrations of calcium carbonate, and in alluvial or outwash deposits where the sand and gravel may consist largely of limestone or dolomite. Hard waters can also be expected where sulfate is abundant, in addition to calcium, either solid in the rocks or dissolved in the water. Most of these areas lie west of the Mississippi River where the climate is arid or semiarid and evaporite deposits are common. In coal-mining areas, concentrations of hardness occur in acid-mine drainage that contaminates streams.

BICARBONATE AND CARBONATE

Bicarbonate (HCO_3) and carbonate (CO_3) are commonly reported as causing alkalinity—the capacity to neutralize acid. Alkalinity, however, is also influenced by hydroxides, iron, silica, and phosphate, all of which react according to temperature, pH, and the concentration of other dissolved solids.

Surface waters commonly contain little or no carbonate, while bicarbonate concentrations may be substantial, approaching 200 mg/L in many cases. Streams that contain no or only minor amounts of bicarbonate may be contaminated by acid. In particular, many of the streams in Appalachia that receive acid drainage from coal mines contain practically no bicarbonate (Biesecker and George, 1966).

Most of the carbonate and bicarbonate in ground water is derived from the atmosphere, the soil, and the solution of carbonate minerals. Carbonate concentrations in ground water generally range from 0 to 10 mg/L, although large concentrations may occur in waters high in sodium. Bicarbonate, on the other hand, is nearly always much higher, ranging from 50 to 400 mg/L; and concentrations in excess of 1000 mg/L are not uncommon, particularly where calcium and magnesium concentrations are low or where sulfate reduction is occurring.

The federal drinking water regulations do not now include alkalinity. The recommended range of alkalinity for public water supplies is 30 mg/L to 400 mg/L (FWPCA, 1968). Concentrations less than 100 mg/L are desirable for domestic water supplies. Alkalinity itself is not detrimental to humans.

Carbonates and bicarbonates, in the form of a soda ash, and sodium bicarbonate (baking soda) are being used successfully for corrosion control in naturally soft water supplies. This is presently true of the supply at Bennington, Vermont, where the naturally soft water has been picking up lead and cadmium from solder in the copper pipes. The same method of treatment is being considered for Seattle, Washington, which also has naturally soft water.

SULFATE

Many sulfate compounds are readily soluble in water. Most of them originate from the oxidation of sulfite ores, the solution of gypsum and anhydrite, the presence of shales, particularly those rich in organic compounds, and the existence of industrial wastes. Sulfate is one of the major dissolved constituents in rain, where it usually appears in concentrations of less than 5 mg/L. The amount is related to natural and human pollution of the atmosphere. This soluble substance may be derived from dust particles; oxidation of SO_2 gas (a major pollutant caused by the burning of high-sulfur fuels) and that of H_2S gas, which is released through the decomposition of organic material; and volcanic discharges.

Sulfur-bearing minerals are common in most sedimentary rocks. In the weathering process, gypsum (calcium sulfate) is dissolved and sulfide minerals are partly oxidized, giving rise to a soluble form of sulfate that is carried away by water. In humid regions, sulfate is readily leached from the zone of weathering by infiltrating waters and surface runoff, but in semiarid and arid regions the soluble salts may accumulate within a few tens of feet of land surface. Where this occurs, sulfate concentrations in shallow ground water may exceed 5000 mg/L, but gradually decrease with depth.

Since sulfate compounds are quite mobile in aquatic systems, there are few circumstances where concentrations will naturally decrease. One exception, however, is sulfate reduction by bacteria. Sulfate-reducing bacteria derive energy from the oxidation of organic compounds, obtaining oxygen from sulfate ions. The bacterial reduction of sulfate produces H_2S gas as a by-product. If sufficient iron is present under moderate reducing conditions, iron sulfides may also be precipitated, thus decreasing the concentration of both iron and sulfate. Much of the H_2S gas escapes directly into the atmosphere if the reduction is by bacteria in the soil.

Sulfates are difficult to remove from water and may produce a detectable taste at concentrations of 300 to 400 mg/L and a medicinal or bitter taste if the concentrations exceed 500 mg/L. Individuals unaccustomed to consuming sulfate-rich (more than 600 mg/L) water soon recognize its cathartic nature, but a tolerance to the laxative reaction can be developed with time. It is mainly for this reason that the National Secondary Drinking Water Regulations (1977) recommended a maximum limit of 250 mg/L of sulfate in drinking water.

Sulfate concentrations range widely in natural waters, from less than 5 mg/L in rain to several thousands in some brines. Streams in humid regions generally contain less than 50 mg/L, although concentrations in excess of 5000 mg/L may occur in streams draining coal or metal-sulfide mines.

Conversely, in arid and semiarid regions sulfate concentrations are considerably higher because the soluble compounds tend to accumulate in the soil and shallow aquifers. Also, many of the sedimentary rocks west of the Mississippi River contain an abundance of sulfate minerals, and subterranean waters migrating through them may accumulate more than 1000 mg/L of sulfate. Figure 5-3 shows the distribution of sulfate in drinking water from ground water sources.

CHLORIDE

Despite the fact that chloride is only a minor constituent of the materials forming the earth's crust, it is a major dissolved substance in some natural waters. In seawater the average chloride concentration is 19 300 mg/L. Rain in coastal areas may contain several tens of milligrams per liter, but inland concentrations generally decrease to less than 1 mg/L. The high concentrations are due to salt crystals originating from sea spray, which serve as nuclei around which moisture accumulates.

In most streams the chloride content is less than either sulfate or bicarbonate. Uncontaminated streams in humid areas generally contain less than 35 mg/L of chloride, although the concentration in some areas is far greater because of contamination by municipal sewage and industrial wastes. In semiarid and arid regions, chloride in streams commonly exceeds 100 mg/L, the higher amounts being due to concentration by evaporation and solution of chloride-rich sedimentary rock.

Chloride in ground water ranges within wide limits (Fig. 5-4). In humid areas it is generally less than 30 mg/L and may exceed 1000 mg/L in arid regions, and brines may contain more than 200 000 mg/L. Chloride moves through earth materials with less retardation or loss than probably any other chemical. For this reason, it has considerable value as a tracer.

Chloride concentrations in ground water may be related to seawater trapped in sediments when they were deposited, it may originate from evaporites, and it is concentrated by evaporation and transpiration, which may precipitate its various salts. Other sources include the solution of dry-atmosphere fallout, particularly in arid regions, and springs, especially hot springs, which provide large concentrations to both sur-

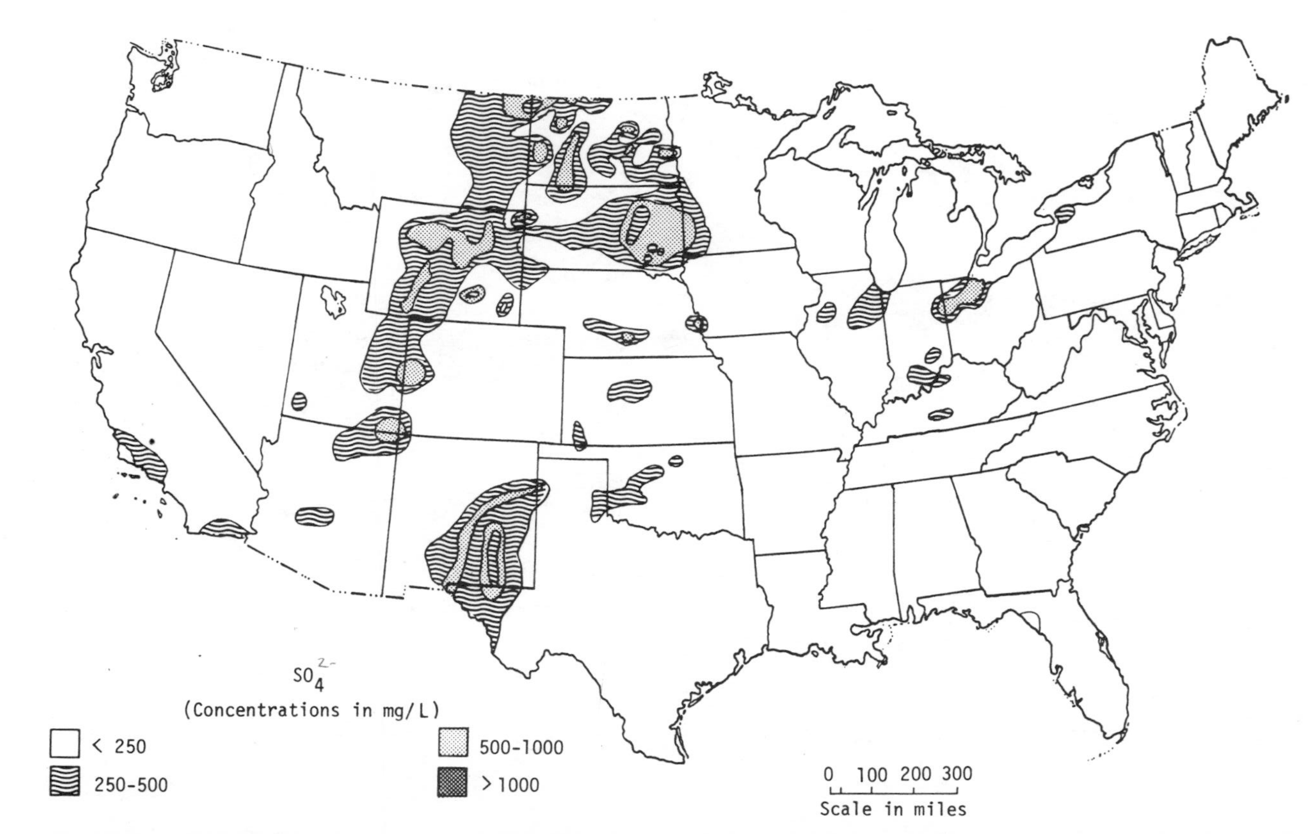

Figure 5-3. Distribution of sulfate content of drinking-water supplies from ground water sources in the coterminous United States.

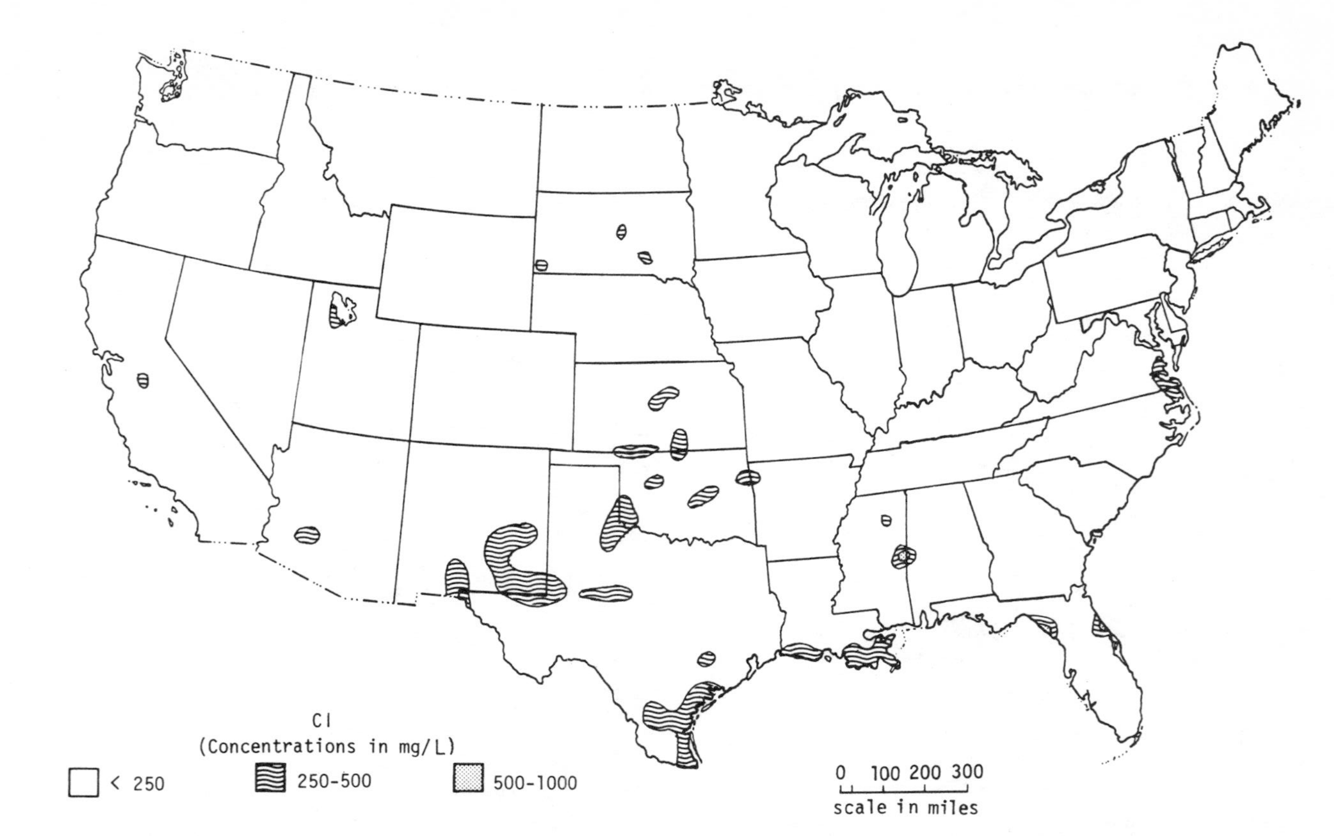

Figure 5-4. Distribution of chloride in drinking-water supplies from ground water sources in the coterminous United States.

face and ground water. Another major source in some areas is pollution. Both human and animal excretions are high in chloride and nitrogenous materials; high concentrations of both may be indicative of pollution.

Concentrations of chloride rarely diminish in aquatic systems because all chloride salts are very soluble. Moreover, the chloride ion is difficult to remove from water supplies because it remains in solution through most of the processes that separate other ions. Since the chloride ion is physically large compared with many other major ions present in water, it may be retained in pore water in clay or shale, while the water itself is transmitted. This may lead to the formation of predominantly sodium chloride brines. Some chloride originating from seawater may remain long in fine-grained marine shales as sodium chloride crystals or solutions.

In water, chloride concentrations in excess of 250 mg/L may impart a salty taste; this is the recommended limit for drinking water supplies. But the threshold of detection, which varies with the individual, is more generally in the neighborhood of 500 mg/L. In excess of 1000 mg/L, it may be physiologically unsafe.

Chlorine gas (Cl_2) readily dissolves in water and is used in treatment as a disinfectant or biocide. Material reacting with the chlorine is destroyed or oxidized, and sufficient chlorine must be used so that some remains as a residual throughout the distribution system.

In some situations, chloride may accelerate corrosion of pipes, boilers, and fixtures. Magnesium chloride–rich water, when heated, may generate HCl, which is highly corrosive.

SILICA (SILICON DIOXIDE)

Silicon is by far the most abundant element in igneous and sedimentary rock. It ranks second only to oxygen as the most abundant element in the earth's crust. Silicon is commonly found in combination with oxygen, with which it forms silica. Despite its abundance, silica concentrations in natural water are generally small, ranging between 1 and 30 mg/L, but concentrations in excess of 100 mg/L are not uncommon (Hem, 1963). Silica may be dissolved from nearly all rocks, but usually only in small amounts. This is because of its low solubility in the normal pH range of natural water. In very acid hot springs or highly alkaline springs, silica concentrations may exceed several hundred milligrams per liter, but large concentrations are the exception.

Silica is widely used in making glass, silicates, ceramics, abrasives, enamels, petroleum products, and metal work, so it may appear in certain waste waters. Silicates are also used in water-treatment systems as coagulants and corrosion inhibitors.

Even after decades of study, the chemistry of silica is still not well known. However, any water sample that contains more than 40 mg/L of silica should be examined in closer detail in order to determine whether pollution is present.

IRON

Iron compounds, very common in rocks and soil, are easily leached by water, particularly by acidic water. Although the earth's crust provides a major source of iron, concentrations in natural water are generally less than 5 mg/L. Water with a low pH usually contains more iron than alkaline water. The general absence of oxygen in ground water leads to higher concentrations of iron in ground water than in streams.

In ground water, iron most commonly occurs in the ferrous state, but it may be quickly and easily oxidized to the ferric state. As oxygen-rich meteoric waters infiltrate, the oxygen is slowly dissipated as it reacts with organic and inorganic oxidizable material in the ground. Below the water table, iron is taken into solution; in a normally alkaline water, it occurs as ferrous [iron (II)] bicarbonate [$Fe(HCO_3)_2$]. This substance is colorless, so that water containing even large concentrations of it will also be clear. When water is removed from a well, however, oxygen in the air, pumping, or distribution systems quickly oxidizes the ferrous (divalent) ion to a ferric (trivalent) state, which hydrolizes to form insoluble ferric [iron (III)] hydroxide [$Fe(OH)_3$]. At this time some of the free carbon dioxide may escape and raise the pH, thus continuing the oxidation process.

The ease with which the reaction can occur becomes evident when we realize that 1 mg/L of ferrous iron requires only 0.14 mg/L of oxygen to oxidize it. It is for this reason that rusty-appearing deposits of iron are so common at springs and at leaking joints in pumps and pipes. Furthermore, ferric hydroxide, which occurs as a precipitate or as particulate matter, is characterized by red and brown colors; as little as 0.3 mg/L, which is readily noticeable, causes staining.

In addition to the iron derived from aquifer minerals, well water may contain iron from corrosion of the metal in pipes, pumps, and fixtures. Because of the near absence of dissolved oxygen, iron concentrations are normally very high in the deeper parts of lakes and reservoirs, particularly during winter when the dissolved-oxygen content is depleted, in streams affected by acid-mine drainage, or in watercourses polluted by wastes from steel mills and metal plants. In areas of acid-mine drainage, the bottoms of many streams are coated with "yellow boy," or iron oxide. Water in the stream, however, may be crystal clear, despite high iron concentrations, if the pH remains low and the ferrous ions are not oxidized.

Crenothrix, Gallionella, and other so-called iron bacteria utilize dissolved iron as a source of energy. Such bacteria may cause rusty water, or "red water," in water supplies. Bacterial colonies also build a slime in water closets, pipes, and pumps; this slime may break loose to flow freely through the distribution system and cause slugs of ferric hydroxide to appear at the faucet.

It is evident that iron occurs as a naturally derived metallic compound, either in the ferrous or ferric state, or it may owe its presence in water supplies to the action of bacteria or other corrosive phenomena. Regardless of the source, iron creates a variety of problems. Concentrations in excess of 0.3 mg/L will cause staining of laundry and utensils. It is usually objectionable for processing food, making beverages, dyeing, bleaching, manufacturing ice, and processing many other items. Potatoes boiled in iron-rich water may turn black; iron combines with tannins in tea and coffee to form a black inky appearance and an objectionable metallic taste; when iron is mixed with whiskey, the solution turns dark or even black. If iron concentrations exceed 0.5 mg/L, wells and well screens are likely to become encrusted and if it exceeds 1 mg/L, coffee becomes unpalatable.

The federal recommended limit of iron in drinking water (0.3 mg/L) is not based on physiological reactions but on aesthetic and taste considerations, namely the stain that occurs at concentrations in excess of 0.3 mg/L, and the metallic taste, the threshold of which may be as low as 0.1 mg/L, depending on the iron compound present.

MANGANESE

In water analyses and reports, iron and manganese may be reported and described together. Although there are many similarities between these elements, a few great differences do exist. Despite the fact that sources of manganese in the earth's crust are far fewer than those of iron, it is present in many natural waters. Manganese-bearing minerals are common in rocks and soils. Manganese may exist in large concentrations in organic material, since it is a plant nutrient. Concentrations of manganese in natural waters are generally 0.02 mg/L or less, although as much as 10 mg/L is not uncommon. The largest amounts of it are found in acidic waters and it begins to precipitate as the water becomes more alkaline. Generally, ground waters contain more iron than manganese. In alkaline ground water, manganese most commonly occurs as manganous bicarbonate.

Not easily oxidized even in surface waters, manganese is more abundant and travels downstream farther than iron. Manganese concentrations greater than 1 mg/L have been reported in streams receiving acid-mine drainage. Excessive concentrations may exist in ground water contaminated by oil-field brine. As is also the case with iron, high concentrations of manganese may be present in deeper parts of lakes and reservoirs where reducing conditions occur, particularly during winter.

Manganese does not appear to have toxicological significance in drinking water, at least in concentrations typical of natural waters. Some reports, however, have described a few examples of manganese poisoning, but apparently these have all occurred under unusual circumstances.

The federal recommended limit for manganese in drinking water, 0.05 mg/L, is based largely on aesthetic and taste considerations. Upon oxidation, manganese in excess of 0.2 mg/L tends to precipitate and form noxious deposits on foods during cooking, and black stains on plumbing fixtures and laundry. Concentrations greater than 0.5 mg/L may impart a metallic taste to both foods and water. As little as 0.1 mg/L of either iron or manganese stimulates the growth of certain bacteria in reservoirs, filters, and distribution lines.

NITRATE

Nitrate in water supplies owes its origin to several possible sources, including the atmosphere, legume plants, plant debris, animal excrement, and sewage, as well as nitrogenous fertilizers and some industrial wastes. Since the atmosphere consists of about 78 percent by volume of nitrogen, some of that present in water originates from this source, but most is generated by the decay of organic matter, and from industrial and agricultural chemicals.

Bacteria decompose organic matter, such as sewage and excrement, and these complex proteins change to ammonia (NH_3), then nitrite (NO_2), and finally nitrate (NO_3). Since the various forms are highly soluble in water, they are easily leached downward from the soil by infiltrating water and may quickly reach the water table. Much of that present in the soil, however, is used by plants, since nitrate is a major nutrient for vegetation and is essential to all forms of life.

In addition to decaying organic matter, fertilizers are a major source of nitrate in water supplies. In several of the heavily fertilized and irrigated regions west of the Mississippi River, the concentration of nitrate in both surface and ground water has increased to alarming amounts.

Nitrate concentrations in water are reported either as nitrate (NO_3) or as nitrate-nitrogen (as N). This system is confusing, so caution is urged when examining an analysis, because the concentration of nitrate reported as NO_3 is considerably different from that reported as nitrate-nitrogen (as N). For example, 45 mg/L of NO_3 is approximately equal to

10 mg/L of nitrate (as N). To convert NO_3 to N, multiply the NO_3 concentration by 0.23; and to convert N to NO_3, multiply the N concentration by 4.43. An understanding of this conversion is essential, because the recommended limit of nitrate (as N) is 10 mg/L; but as NO_3, it is 45 mg/L.

Nitrate (as N) in concentrations greater than 10 mg/L has been known to cause infant methemoglobinemia, a disease characterized by cyanosis, a bluish coloration of the skin. This disease may occur when a nursing child consumes either a formula or milk directly from its mother that contains large concentrations of nitrate. Some evidence indicates that high concentrations of nitrate in drinking water for livestock has resulted in abnormally high mortality rates in baby pigs and calves and in abortion instances in brood animals.

Shallow ground water supplies are susceptible to pollution by nitrates. Shallow dug or bored wells, particularly, may contain excessive nitrate if they are improperly constructed so that surface water is allowed to flow into them or if they are near septic tanks, cesspools, or barnyard wastes. In cases such as these, the supply should be abandoned, because the pollutants might contain pathogenic organisms as well as other undesirable organic substances.

If polluted by human excrement, the water supply might contain relatively large concentrations of chloride in addition to nitrates, and the presence of both provides a warning that the homeowner should have the supply checked for bacterial contamination. Furthermore, the concentration of nitrate in ground water can fluctuate greatly from one season to the next. During the wet season, particularly while plants are still dormant, infiltrating rain or snowmelt may leach large amounts of nitrate to the water table. Most of it is either used by plants during the growing season or remains in the ground because of a soil-moisture deficiency, but after crops are harvested in the fall, nitrate may again be leached down to pollute shallow aquifers. In view of the cyclic nature of nitrate in shallow ground-water supplies, judgment must be used so that sampling is accomplished at the appropriate time.

FLUORIDE

Fluoride occurs in just a few types of rocks and is only slightly soluble in water. High concentrations are relatively rare, being generally less than 0.5 mg/L. Higher concentrations do occur in aquifers in a few areas; and certain insecticides, chemical wastes, and airborne particles and gases from aluminum smelting plants can pollute both surface and ground water, as well as the soil.

The element fluoride is utilized by animals, including humans, in the structure of bones and teeth. Its effect in reducing the formation of dental caries in children has already been mentioned. Large concentrations of fluoride are toxic, but Smith (1944) reports that adults may safely drink 2 gal (7.57 L) of water each day containing 10 mg/L of fluoride. Even when daily intakes are 15 to 20 mg/L, several years are required to induce chronic fluorosis in an adult. Unquestionably, the addition of fluoride to drinking-water supplies has been of tremendous benefit to the nation's population.

The National Primary Drinking Water Standards limit the amount of fluoride in drinking water on the basis of the annual average of the maximum daily air temperature (Table 5-4). The concentration ranges from a low of 1.4 mg/L in hot areas to a maximum of 2.4 mg/L in cold regions. The limit is based on air temperature because it is assumed that children in warm or hot areas will drink more water than will those in colder climates.

pH

The pH of a solution describes its hydrogen-ion activity. The pH scale ranges between 0 and 14 (Fig. 5-5). Water with a pH of 7 is neutral and less than 7 is acidic; a pH greater than 7 denotes a basic solution. Dissolved gases, such as carbon dioxide and hydrogen sulfide, and ammonia strongly influence the pH of a solution.

The pH of a water sample can change drastically and rapidly with time, especially when the sample is exposed to air, changes in temperature, biologic activity, and other phenomena. For this reason, the pH of a water sample should be determined as soon as it is collected.

Because carbonates are so abundant in nature, most waters are slightly basic. The pH of most streams and ground water supplies ranges between 6 and 8.5. Water contaminated by acid mine drainage and by certain industrial wastes, as well as water found in a few highly mineralized hot springs, may be strongly acidic.

Table 5-4. Maximum Concentrations of Fluoride Permitted in Drinking Water

Temperature		
°F	°C	*Level mg/L*
53.7 and below	12.0 and below	2.4
53.8 to 58.3	12.1 to 14.6	2.2
58.4 to 63.8	14.7 to 17.6	2.0
63.9 to 70.6	17.7 to 21.4	1.8
70.7 to 79.2	21.5 to 26.2	1.6
79.3 to 90.5	26.3 to 32.5	1.4

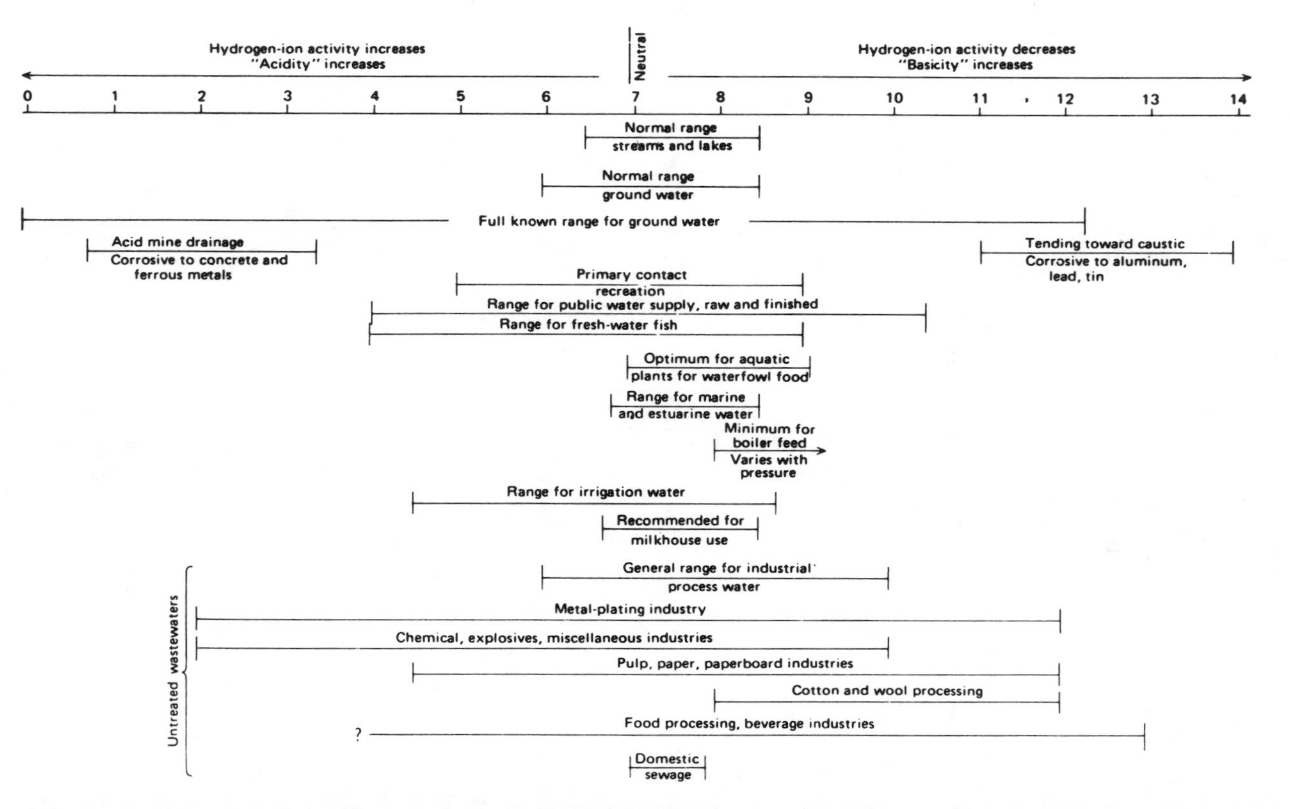

Figure 5-5. pH ranges in relation to use. [References: White, Hem, and Waring (1963), FWPCA (1968), Hem (1970), Rudolfs (1953), Ciaccio (1971), and Comp (1963).] (*From U.S. Geological Survey.*)

The pH of some surface waters may be characterized by annual or even diurnal fluctuations. For example, a stream may have a higher pH in summer than winter because of active summertime photosynthesis of aquatic plants. This process depletes the dissolved carbon-dioxide content and raises the pH. Similarly, on a daily basis, plants use carbon dioxide during daylight hours, thus raising the pH; but at night the increase in carbon dioxide lowers it.

The pH of drinking water, in itself, has no effect on health. On the other hand, corrosion is associated with pH levels less than 6.5. Corrosion releases metals, such as lead, zinc, copper, and cadmium from pipes and plumbing fixtures, and these substances can be toxic. Furthermore, if the pH is less than 4, the water may have a sour taste.

Other problems can occur in water with a higher pH. For example, when the pH exceeds 8.5, water may have an alkali taste, scale may form in pipes and equipment, the germicidal activity of chlorine is reduced, and trihalomethane formation is accelerated.

GASES

Gases are important in water supplies only when they are present in concentrations sufficient to create an explosion hazard or an undesirable taste and odor or to be corrosive. Two types of gases commonly present are hydrogen sulfide (H_2S) and methane (CH_4).

Hydrogen sulfide is a flammable, poisonous gas, highly soluble in water, that has a characteristic rotten-egg odor. Hydrogen sulfide originates from the decomposition of organic matter, including sewage and certain industrial wastes. It is present in many ground water supplies because of anaerobic decomposition of naturally occurring organic deposits. Its presence promotes the growth of certain bacteria that may clog pipes and well screens.

The U.S. Environmental Protection Agency (EPA) has recommended a maximum hydrogen sulfide concentration in drinking water of 0.05 mg/L, which is the minimal amount detectable by taste. In many places, however, hydrogen sulfide–rich waters are prescribed or recommended for respiratory, metabolic, and skin disorders. This gas forms the basis of many health spas, particularly in Europe.

Methane is a colorless, tasteless, odorless, lighter-than-air gas that can present an explosion hazard. Because of the danger of asphyxiation and its explosive nature, water supplies that contain methane should be carefully vented so that it cannot accumulate in distribution lines, water heaters, pressure tanks, water-treatment equipment, or well houses.

Methane is a product of anaerobic decomposition of organic material. It may originate in certain industrial wastes, particularly in natural gas and petroleum refineries, but in ground water it generally is derived from the decomposition of naturally occurring organic matter.

Other gases commonly present in water include dissolved oxygen and carbon dioxide. Both are undesirable because of their relation to corrosivity.

Corrosion is a form of chemical erosion that decreases the useful life of water systems, reduces their structural strength, and increases the mineral content of the water. When corrosion occurs in water-distribution or pumping equipment, the water supply may have an undesirable taste or odor and red water may be common. Although the various chemical, electrical, and biological causes of corrosion are still not well understood, it generally occurs in iron pipes that transport water containing high concentrations of dissolved oxygen and carbon dioxide, coupled with low alkalinity and pH. High concentrations of hydrogen sulfide, dissolved solids, and chloride, as well as increasing water temperature may also increase the rate of corrosion. Other contributing factors may be the presence of sulfur dioxide or similar gases, organic acids, and iron sulfate.

Dissolved oxygen can greatly accelerate corrosion in acidic, neutral, or even slightly alkaline waters. High concentrations of dissolved oxygen are expected in streams, lakes, and ponds, but the concentration is usually near zero in deep aquifers. Carbon dioxide in excess of 50 mg/L, even in the absence of dissolved oxygen, accelerates corrosion, as does chloride in excess of 300 mg/L when it occurs in acidic water.

In addition to direct chemical corrosion, other types include dezincification (removal of zinc), galvanic (caused by a connection of two different metals that set up an electric current), and bacterial. In a number of instances sulfate-reducing bacteria have been found to be the major causes of corrosion damage (Postgate, 1963). Moreover, iron bacteria may also cause pronounced corrosion.

The proposed Secondary Drinking Water Regulations recommended that water used for consumption should be noncorrosive. The major reasons are related to aesthetics, health, and economics. Corrosive water may contain not only staining substances, but also toxic concentrations of cadmium and lead (dissolved from pipes and solder), and these waters reduce the life expectancy of distribution systems. Unfortunately, rapid, simple methods are not presently available for measuring the corrosivity of water.

TRACE ELEMENTS

The trace elements form a group of generally poorly understood substances that normally occur in concentrations that are less than 1 mg/L in water. Many trace elements are essential to human well-being, but in some cases there is only a small safety range between requirement and toxicity.

Cobalt, for example, is a component of vitamin B_{12}, which is essential to life. The lack of minute amounts of copper in the diet results in nutritional anemia in infants, but large concentrations may cause liver damage. Even minor amounts of substances, such as arsenic, lead, cadmium, mercury, and selenium, may cause serious adverse physiological effects in human beings.

Fortunately, the rocks and minerals that are sources of trace elements are only slightly soluble and rarely occur in large amounts. Therefore, abnormally large concentrations of trace elements of natural origin in water are infrequent. On the other hand, inadequate disposal of certain industrial and mining wastes, careless handling of some agricultural sprays, and use of several common household items can pollute water supplies with dangerous quantities of toxic substances. In addition, corrosive water may release lead, copper, and cadmium directly into drinking-water supplies.

Health problems related to an excessive intake of trace elements are probably not widespread, but the symptoms of the diseases they may cause are sometimes so subtle or commonplace, at least in the early stages, that they are not easy to diagnose. It is because of this potential danger to health that federal drinking-water standards include several trace elements, such as arsenic, barium, cadmium, chromium, lead, mercury, selenium, and silver.

Arsenic is a well-known poison that occurs naturally in many rocks, minerals, and soils. Several industrial processes involved with ceramics, tanneries, chemicals, and metal preparation require its use, but the manufacture of pesticides consumes the largest amount. The presence of arsenic in water supplies is not solely the result of pollution. Naturally occurring arsenic compounds have been reported in waters from the western part of the United States.

The toxicity of arsenic to humans is well known. It accumulates in the body and causes arsenosis. The effects of the poison, when ingested in small amounts, appear very slowly; it may take several years for the poisoning to become apparent. Chronic arsenosis, in its most extreme form, causes death. Arsenic may be carcinogenic and is known to affect the liver and heart. The federal limit of arsenic in drinking water is only 0.05 mg/L.

Barium is quickly precipitated in natural water systems or removed by sedimentation and absorption and, therefore, generally occurs in barely detectable concentrations. Although harmful concentrations do appear in some mineralized springs and where barium compounds are mined, most pollution is probably related to the disposal of wastes from some special alloy plants, in the paint industry and at some ceramic and glass plants.

Federal drinking-water regulations limit barium concentrations in drinking water to 1 mg/L because of its possible toxic effects on the heart, blood vessels, and nerves. Apparently barium does not accumulate in bone, muscle, kidney, or other tissue, so there is no danger of a cumulative effect.

Only under very unusual circumstances does cadmium appear in measurable concentrations in natural water supplies, but waters contaminated by certain industrial wastes may contain several milligrams per liter. Cadmium salts are present in industrial wastes, particularly those from electroplating and chemical industries, and milling and mining wastes from lead-zinc mines.

Cadmium collects in the liver, kidneys, pancreas, and thyroid of humans and other animals. Like arsenic, once in the body it tends to accumulate and is likely to remain there. Cadmium is also suspected of acting as at least one of the major causes of "itai-itai" (ouch-ouch), a serious, painful bone disease found in some parts of Japan. For these reasons, federal standards list a maximum limit of 0.01 mg/L of cadmium in drinking water.

Chromium is neither beneficial nor essential to the body and does not accumulate in it. Federal standards limit hexavalent chromium in drinking-water supplies to a maximum of 0.05 mg/L. This was the lowest amount analytically determinable when this substance first appeared in the U.S. Public Health Service recommended limits in 1946. Although when inhaled chromium can be carcinogenic, test results seem to indicate that the hexavalent form, even when consumed by test animals for more than a year at concentrations of 0.45 to 25 mg/L, does not necessarily show evidence of toxic response. The 0.05 mg/L limit, however, must remain in force until additional research is carried out.

Hexavalent chromium occurs in only minute concentrations in unpolluted water supplies. It may appear at considerably higher concentrations in water polluted by certain industrial wastes, such as those related to electroplating processes, aluminum anodizing operations, paints, dyes, explosives, ceramics, and paper production. Chromium is also used as a corrosion inhibitor and may appear in high concentrations in streams that receive chromium-treated cooling waters.

Federal drinking-water standards impose a limit of 1.0 mg/L of copper, not because it is a health hazard, but because it can impart an undesirable taste. Copper is both essential and beneficial and its lack can lead to nutritional anemia in very young children; adults require about 3 mg each day. For the most part, copper is readily passed from the body in waste, although it does tend to accumulate in the liver.

Concentrations of copper in water in excess of about 0.05 mg/L are generally the result of pollution, either from industrial or mining wastes or the corrosion of copper plumbing. In some cases, it may originate from copper sulfate, which is used to control undesirable plankton in lakes, ponds, and reservoirs.

Lead is another well-known poison that accumulates within the body. Fortunately, lead concentrations normally present in water supplies are very small. In fact, they are generally barely detectable. One source of lead in drinking water, far more prevalent in the past than in the present, thanks to the now widespread home use of copper tubing, is the corrosion of lead pipes. The federal limit on lead in drinking water is 0.05 mg/L.

Several common lead compounds may be sources of this toxic heavy metal in water supplies. Lead acetate is used in the printing and dyeing industries. Lead chloride is used in the manufacture of lead-base paints and in solder. The high concentration of lead in some sediment samples below some effluent outfalls near oil refineries is probably the result of tetraethyllead, which is also produced by the burning of gasoline.

In the late 1960s mercury was discovered in sediment, fish, and water in Lake Erie. The toxic effects of this heavy metal alerted health officials to a distinct health hazard. In this particular case, the real culprit was organic mercury in the form of methylmercury. Because of the toxic nature of mercury, the federal government established a limit for public water supplies of 0.002 mg/L of mercury.

Measurable concentrations of naturally occurring mercuric salts are not common in water. However, disposal of medicinal products, disinfectants, pigments, and materials used in photoengraving may lead to serious local problems.

In some respects, selenium is chemically similar to sulfur, although much less common. In the 1930s selenium was discovered to be the cause of a livestock disease in the western United States. The main source of the selenium was vegetation, but local drainage also contained relatively large amounts. Because of the potentially toxic effects of selenium, the federal government established a limit of 0.01 mg/L on its concentration in drinking water.

Although retained in the liver and kidney in small amounts, selenium salts, for the most part, are excreted. In trace quantities, selenium appears to be essential for nutrition of human beings and other animals, but larger concentrations produce definite toxic symptoms. In fact, some surveys have shown that the rate of dental caries on permanent teeth is higher in seleniferous areas than in nonseleniferous regions. The limit on selenium in water supplies is based on the similarity between arsenic and selenium poisoning, its dental effects, and its known toxicity to livestock (McKee and Wolfe, 1963).

Although not a widespread problem, water supplies contaminated by wastes from paints, dye and glass production, electrical apparatus, the rubber industry, and certain insecticides could cause a local danger of selenium poisoning.

Silver may be leached into water sources from various ores; but, since many of its salts are insoluble, it cannot be expected to occur in significant amounts. Silver can occur in waste waters from electroplating, certain food processing and beverage industries, and from photography. Silver oxide has been used for water purification and its bactericidal power is well known.

The chronic intake of silver salts, which was particularly common in the early part of the twentieth century, produced a permanent bluish skin color called argyria. A limitation of silver in water supplies was first initiated by the U.S. Public Health Service in 1962. According to McKee and Wolfe (1963), the limitation apparently was established not to avoid health hazards but to prevent water-treatment-plant operations from adding excessive concentrations to the supply for disinfection. The present limit of silver in public water supplies is 0.05 mg/L.

Zinc salts are used in galvanizing and in manufacturing paint pig-

Table 5-5. Finished Water—Before and After Passage through a Home Water Softener, μg/L

			After 24-h contact with copper tubing	
Element	*Unsoftened*	*Softened*	*Unsoftened*	*Softened*
Zinc	<15	<15	290	100
Cadmium	<15	<15	<15	<15
Arsenic	<70	<70	<70	<70
Boron	140	140	77	137
Phosphorus	<70	<70	<70	<70
Iron	79	< 8	77	< 8
Molybdenum	<30	<30	<30	<30
Manganese	< 7.5	< 7.5	< 7.5	< 7.5
Aluminum	<30	<30	<30	<30
Beryllium	< 0.15	< 0.15	< 0.15	< 0.15
Copper	37	< 8	245	126
Silver	< 1.5	< 1.5	< 1.5	< 1.5
Nickel	<15	<15	<15	<15
Cobalt	<15	<15	<15	<15
Lead	<30	<30	3780	2250
Chromium	< 8	< 8	< 8	< 8
Vanadium	<30	<30	<30	<30
Barium	29	< 2	2	< 2
Strontium	93	< 2	129	< 2

Source: Kopp, 1969

ments, pharmaceuticals, cosmetics, and several insecticides. The solubility of many of these salts in water accounts for their presence in industrial wastes.

Except at very high concentrations, zinc alone has no known adverse physiological effects upon human beings. In fact, it is both essential and beneficial for human nutrition. Some evidence indicates that it aids in healing wounds. Zinc is not stored in the body and very little is known about its deficiency in human beings.

Large concentrations of zinc in drinking water are undesirable for several reasons. Zinc produces an objectionable taste and may cause water to appear milky or, upon boiling, to seem to have a greasy surface scum. It is largely because of taste considerations that the federal drinking water standards impose a limit of 5.0 mg/L.

The concentrations of many elements found in water are greatly reduced after passage through a water softener (Table 5-5). Particularly large reductions of iron, copper, barium, and strontium occur. Concentrations of many substances also increase due to corrosion after contact with copper tubing for 24 h (Table 5-5). On the other hand, if a water tap is allowed to flow to waste for a minute or so, the concentration of the various constituents derived by corrosion are reduced to barely detectable amounts.

ODOR

As a general rule, individuals have a more acute sense of smell than taste. In many instances minute concentrations of some substances are more readily detectable by odor than by several analytical methods. Disagreeable odors are caused by a great variety of materials, particularly living micro-organisms, decaying organic matter, sewage, and some industrial wastes. Probably the most common odors in water are the rotten-egg smell of hydrogen sulfide, the gasoline smell in supplies polluted by certain hydrocarbons, and the pungent odor of chlorine. The latter may result from reactions between organic matter and chlorine during water treatment. These substances combine to form chlorophenols, which have an extremely low taste and odor threshold.

Federal standards limit odor in water supplies to a threshold odor number 3. This sytem is based on the amount of sample in its most diluted concentration that produces a perceptible odor. Odors are generally removed with activated carbon or by aeration.

ORGANISMS

Pathogenic (disease-producing) organisms that can occur in water supplies range from the ultrasmall virus to the microscopic bacteria and finally to relatively large cysts.

The relatively large ameba called *Entamoeba histolytica* is one of the most important parasites found in humans. It is able to survive for a few days outside a human's body (colon) because it can form a tough covering called a *cyst*. Water polluted with fecal wastes containing these cysts may, when consumed, cause a disease known as *amebic dysentery*. Because of their large size, these cysts are easily removed from water by filtration, but even treated water may become polluted through carelessness. Defective plumbing, for example, was the cause of a severe outbreak during the World's Fair in Chicago in 1933, and caused nearly 1000 cases of amebiasis and 58 deaths.

Giardia lamblia is a protozoan responsible for giardiasis, the number one waterborne disease in the United States. It causes acute diarrhea, often lasting 2 to 3 months. From 1971 to 1974, 14 known waterborne outbreaks of giardiasis were documented in the United States, affecting more than 5000 individuals. Most of the outbreaks were related to the consumption of untreated surface or ground water or surface water whose treatment was only disinfection. *Giardia* cysts are not destroyed by normal chlorine concentrations and contact time, but are easily removed by fine filtration.

The fecal coliform bacteria and the coli-aerogenes group, although not pathogenic themselves, are indicators of fecal pollution. They live in great abundance in the intestines of warm-blooded animals and are eliminated, also in great abundance, with feces. Their presence in a water supply indicates recent and possibly dangerous pollution. In addition to coliforms, many other bacteria and viruses also inhabit the intestine, including fecal streptococci and enteric viruses, both of which are pathogenic. All three groups may exist in an individual, but the coliform group is much more abundant and easier to grow and identify. Therefore, it is used as an indicator of pollution and, when present, suggests that pathogenic forms might also be present.

The most common waterborne diseases caused by bacteria include typhoid, paratyphoid, Asiatic cholera, and bacterial dysentery. Other diseases that may be transmitted by water include tularemia, brucellosis, shigellosis, infectious hepatitis, and Weil's disease (jaundice). Even the anthrax spore can be carried on hair floating or suspended in water. In any case, most waterborne diseases caused by bacterial agents are the result of pollution. Diseases may be transmitted by consuming (1) untreated water that contains fecal matter; (2) treated water that has been polluted because of inadequate protection against cross connections, or unsafe or unsanitary plumbing; and (3) inadequately treated water as a result of faulty equipment, lack of maintenance, or improperly trained operators.

Some 76 viruses are pathogenic to humans and those that occur in polluted water include the adenoviruses, Coxsackie and ECHO viruses, reoviruses, polioviruses, and those that cause infectious hepatitis. Contrary to popular belief, viruses can migrate considerable distances in water and even in the ground, particularly in limestone regions where sinkholes and underground caverns occur.

During the period from 1971 to 1974 there were 13 reported outbreaks of waterborne viral hepatitis, affecting at least 351 people in the United States (Craun, McCabe, and Hughes, 1976). Some 66 outbreaks have been reported since 1946; of 22 outbreaks occurring in municipal treatment systems, 14 were related to contaminated water-distribution systems, and inadequate or interrupted disinfection.

REFERENCES

Biesecker, J. E., and J. R. George, 1966: "Stream Quality in Appalachia as Related to Coal-Mine Drainage, 1965," U.S. Geological Survey Circular 526.

Craun, G. F., L. J. McCabe, and J. M. Hughes, 1976: Waterborne Disease Outbreaks in the U.S.—1971–74, *J. A.W.W.A.*, vol. 68, pp. 420–424.

Durfor, C. N., and Edith Becker, 1962: *Public Water Supplies of the 100 Largest Cities in the United States*, U.S. Geological Survey Water-Supply Paper 1812.

Federal Water Pollution Control Administration (FWPCA), 1968: Report of the Committee on Water Quality Criteria.

Hem, J. D., 1963: *Study and Interpretation of the Chemical Characteristics of Natural Water*, U.S. Geological Survey Water-Supply Paper 1473.

Kopp, J. F., 1964: "The Occurrence of Trace Elements in Water," in *Proc. 3rd Annual Conference, Trace Elements in Environmental Health III*, D. D. Hemphill (ed.), University of Missouri, Columbia, pp. 59–73.

McKee, J. E., and H. W. Wolfe, 1963: "Water Quality Criteria," California State Water Quality Control Board Publ. 3-A.

National Interim Primary Drinking Water Regulations, U.S. Environmental Protection Agency, EPA-570/9-76-003.

Office of Saline Water, 1969: "Communities of Over 1000 Population with Water Containing in Excess of 1000 ppm of Total Dissolved Solids," Office of Saline Water, Research and Development Progress Report 462.

Postgate, J. R., 1963: Versatile Medium for Enumeration of Sulfate-Reducing Bacteria, *Appl. Microbiol. J.*, vol. 11.

Rainwater, F. H., and L. L. Thatcher, 1960: *Methods for Collection and Analysis of Water Samples*, U.S. Geological Survey Water-Supply Paper 1454.

Smith, O. M., 1944: The Detection of Poisons in Public Water Supplies, *Water Works Eng.*, vol. 97.

6

Sampling and Analyzing Water

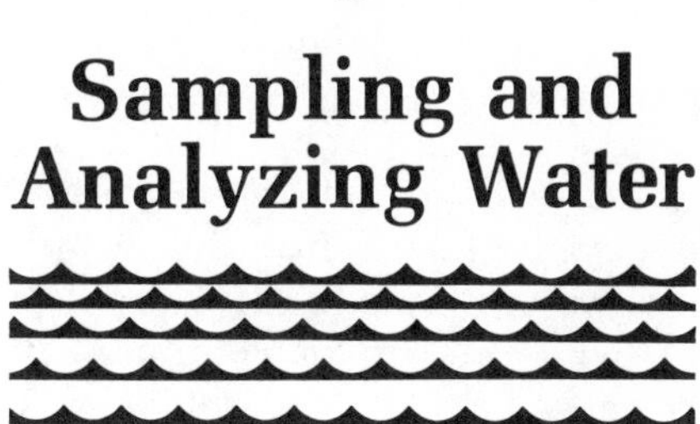

A new home represents one of the most significant investments a person will make during his or her lifetime. Consequently great care is usually taken in selecting a home large enough to meet the needs of the family and tasteful enough to aesthetically please the new residents. Unfortunately, one of the most important selection criteria is often overlooked: the availability of good-quality water for drinking and the sanitary and household needs of the family.

Too often new homeowners discover problems associated with their domestic water supply only after they have already signed the purchase papers or moved into the house. Because the water supply was not considered before the home was purchased, the initial reaction is usually extreme disappointment. Fortunately, after homeowners examine the problem, they discover that nearly all water-quality problems can be alleviated by or eliminated with modern domestic water-treatment equipment.

This chapter underscores the importance of water sampling and water analysis. There are other reasons for evaluation of water quality, however, and these reasons will be examined as well.

WHY SAMPLE AND ANALYZE WATER FOR A DOMESTIC SUPPLY?

The primary requirement of any water supply is that it be free of biological or chemical contaminants that may threaten the well-being of humans, pets, livestock, or vegetation. Yet establishing the safety of a water source is not a simple process, and pollution can result from natural causes. The purity of a water supply cannot be judged by the clarity of the water at the tap or at its source. Picturesque mountain streams are not exempt from natural or human pollution; the stories regarding the safety of spring and well water are nothing more than stories. Although water that has moved underground has undergone a degree of natural filtration, its purity cannot be guaranteed. In fact, dissolved minerals from subsurface materials can make some spring or well water unsafe to drink, or at least aesthetically displeasing.

Another reason for sampling a domestic water supply is to ascertain the cause of a specific problem or complaint. Homeowners often complain that their water has an unusual or unpleasant taste, odor, or color; stains fixtures and clothing; contains sediment; corrodes or encrusts plumbing; or causes illness to family or guests.

The process of sampling and analysis of domestic water supplies not only evaluates the cause of a particular problem but also determines the kind of treatment and the size of the equipment necessary to provide clear, safe, palatable water in desired quantities. In some localities, local, county, or state health ordinances may require domestic-water-supply systems to meet certain minimum standards for physical, chemical, or biological constituents. Therefore, sampling and analysis of water becomes a prerequisite for meeting local laws.

If a new home is constructed where alternative water supplies are available, the homeowner may wish to obtain samples from the different sources to ascertain which provides the best water for domestic use, or which requires the least amount of treatment.

SAFETY OF THE WATER SUPPLY

The phrase "good-quality water" is a widely used expression with many different meanings. Any definition of good-quality water includes the bias of the individual consumer and the intended use of the water. Thus, water suitable for washing cars or watering lawns may be totally unacceptable to a dish washer because it spots glassware and silverware. However, the meaning of a safe-water supply is much more specific. Safe-water supply refers to water free from pathogenic, or disease-causing, organisms and from minerals and organic substances that can have adverse physiological effects. In addition, the water should be aesthetically acceptable. For example, it should be free from apparent turbidity, color, odor, or objectionable taste. In brief, a safe water supply refers to water that can be consumed with pleasure in any desired amount without concern for adverse health effects.

Sanitary Survey

Whether selecting a source for a new water supply or testing an existing one, the importance of a sanitary survey of the water source cannot be overemphasized. The survey locates potential sources of water-quality degradation in and around the point of water withdrawal and determines whether special chemical or biological tests are needed to identify potential sources of pollution.

A sanitary survey entails a visual examination of the surface locality from which the water is to be withdrawn. On-site and neighboring waste-disposal systems, both functioning and abandoned, are located and their potential for contaminating surface- and ground-water supplies is determined. The building contractor, the home's previous owner, or neighbors usually know where nearby septic tanks are located. If not, a metal detector can be used to find the metal handle on the cover of the septic tank. In addition, nearby sewer lines, surface and underground fuel-storage tanks, chemical and industrial plants, graveyards, and landfills represent potential sources of pollution to be identified and noted.

Most counties or municipalities keep records of their sewer lines. The local public works department can usually supply a map of the sewer lines in a neighborhood. Also, a sewer line's location can often be determined by drawing a straight line between two manhole covers.

Nearby gasoline stations may present another potential source of pollution. Although many stations now store gasoline in corrosion-resistant fiberglass tanks, older service stations used steel tanks that are susceptible to corrosion and subsequent leakage.

Nearby chemical or industrial plants may have, in the past, disposed of their wastes through evaporation ponds, liquid-waste injection wells, or onsite burial of solid wastes. Such practices have frequently resulted in regional ground and surface-water pollution. Although recent federal and state regulations have restricted such activities, they are still common in some areas. Even where such practices have ceased, existing buried wastes may still be slowly leaching into the local ground water systems.

Sanitary landfill sites and town dumps have always represented a threat to regional ground water supplies. As with industrial waste disposal, federal and state regulations are reducing the contamination potential of solid-waste-disposal sites. Solid-waste disposal will continue to represent a potential hazard, however, to surface- and ground water supplies for many years to come.

Because of the need to detect all health hazards and assess their present and future importance, persons trained and competent in public-health engineering and the epidemiology of waterborne diseases should conduct the sanitary survey. For an existing supply, the sanitary survey should be made at a frequency compatible with the control of health hazards and the maintenance of a good sanitary quality. Afterward, the information furnished by the sanitary survey should be forwarded to the laboratory that will be analyzing the water sample.

The following outline covers the essential factors which should be investigated or considered in a sanitary survey. Not all the items are pertinent to surveying any one supply.

Ground Water Supplies

1. Character of local geology; slope of ground surface.
2. Nature of soil and underlying porous strata; whether clay, sand, gravel, rock (especially porous limestone); coarseness of sand or gravel; thickness of water-bearing stratum; depth to water table; location, log and construction details of local well, whether operating or abandoned.
3. Slope of water table, preferably as determined from observation wells or as indicated, presumptively but not certainly, by the slope of ground surface.
4. Extent of drainage area likely to contribute water to the supply.
5. Nature of and distance and direction to local sources of pollution.
6. Possibility of surface-drainage water entering the supply and of wells becoming flooded; methods of protection.
7. Methods used for protecting the supply against pollution from sewage-treatment and waste-disposal sites.
8. Well construction:
   a. Total depth of well.
   b. Casing: diameter, wall thickness, material, and length from surface.
   c. Screen or perforations: diameter, material, construction, locations, and lengths.
   d. Formation seal: material (cement, sand bentonite, etc.), depth intervals, annular thickness, and method of emplacement.
9. Protection of well at surface: presence of sanitary well seal; height that casing projects above ground, floor, or flood level; protection of well from erosion and animals.
10. Pumphouse construction (floors, drains, etc.); capacity of pumps; drawdown when pumps are in operation.

11. Availability of an unsafe supply, usable in place of a normal supply, and hence representing a threat to the public health.

12. Disinfection: equipment, supervision, test kits, or other types of laboratory control.

Surface-Water Supplies

1. Nature of surface geology: character of soils and rocks.
2. Character of vegetation, forests, and cultivated and irrigated land, including salinity, effect of irrigation water, etc.
3. Population and sewered population per square mile of catchment area.
4. Methods of sewage disposal, whether by diversion from watershed or by treatment.
5. Character and efficiency of sewage-treatment works located in watershed area.
6. Proximity of sources of fecal pollution to intake of water supply.
7. Proximity, sources, and character of industrial wastes, oil-field brines, acid-mine waters, etc.
8. Adequacy of quantity of supply.
9. For lake or reservoir supplies: wind direction and velocity data, drift of pollution, sunshine data (algae), evaporation data.
10. Character and quality of raw water: coliform organisms (MPN), algae, turbidity, color, objectionable mineral constituents.
11. Nominal period of detention in reservoir or storage basin.
12. Probable minimum time required for water to flow from sources of pollution to reservoir and through reservoir intake.
13. Shape of reservoir, with reference to possible water currents induced by wind or reservoir discharge, from inlet to water-supply intake.

14. Protective measures taken to control fishing, boating, landing of airplanes, swimming, wading, ice cutting, animals on marginal shore areas and in or upon the water, etc., in watershed areas.
15. Efficiency and constancy of policing.
16. Treatment of water: kind and adequacy of equipment; duplication of parts; effectiveness of treatment; adequacy of supervision and testing; contact period after disinfection; free chlorine residuals carried.
17. Pumping facilities: pumphouse, pump capacity and standby units, storage facilities.

BIOLOGICAL SAFETY OF WATER

Waterborne Diseases

To many homeowners, the most important consideration about any water supply is its freedom from disease-producing organisms. Regardless of its source, water is susceptible to biological contamination at any time. For this reason, it is necessary to sample water periodically and to have it analyzed for bacteriological contamination.

Chapter 1 described examples of waterborne diseases contaminating wells. Basically, there are three types of pathogenic (disease-producing) organisms that can affect the safety of water: bacteria, protozoa, and viruses. Bacteriologic and protozoic pathogens are known to cause typhoid, dysentery, cholera, and some types of gastroenteritis. Viruses account for more than 100 human enteric maladies including polio, infectious hepatitis, and some forms of gastroenteritis.

A bacteriological analysis of a water sample will show the disease-producing organisms present at the time of the sampling. Since there are a large and varied number of pathogens, no single sample of contaminated water will contain more than a few varieties, so detection of pathogens by routine water analysis is extremely difficult.

Because speed and accuracy are so essential, laboratory scientists prefer to utilize an organism that can serve as a reliable measure of contamination. The indicator organism should satisfy the following requirements:

1. To be a reliable measure of contamination, an organism must indicate the potential presence of specific pathogenic organisms in either natural or treated water. Such an organism must react in a manner similar to that of the pathogenic organisms in both the natural water supply and the treated water.

2. The indicator organism must be present in greater numbers than the pathogenic organism; otherwise, the pathogenic organism itself would serve the purpose.
3. The indicator organism must be readily identifiable through simple analytical tests.
4. The quantity of indicator organisms in the water must indicate the degree of contamination.

Three types of bacteria or groups of bacteria are used to measure bacteriological contamination potential: fecal coliforms, fecal streptococcus, and total coliform. Presence of members of the fecal coliform group indicates fecal contamination from warm-blooded animals. This group is readily identified by its ability to ferment lactose, producing gas at 44.5°C. Its presence usually indicates that a water supply may be contaminated by liquid wastes from a septic tank or sewer line.

Presence of the fecal streptococcal group indicates fecal contamination from humans or chickens. The variety of fecal streptococcus present often indicates the cause of contamination. The third group, total coliform, is the traditional indicator of the suitability of a particular water for domestic, drinking, cooking, or other uses. The cultural reactions and characteristics of this group have been studied extensively and are described in many texts on bacteriology of water and sanitation.

Coliform bacteria are usually harmless. Whenever a sample of water shows the presence of coliforms, however, the water is immediately assumed to contain human sewage and is therefore considered unsafe to drink. It is true that water can contain coliform bacteria and still not contain any pathogenic organisms, but it is certainly unwise to assume that the people drinking the water supply will remain healthy.

A water analysis resulting in a positive indication of coliform bacteria implies a potential danger. (*Note:* A positive test can also result from contamination of the bottle or the sample of water.) A negative result from one test, however, means only that dangerous bacteria were probably not present *at the time* the sample was taken. This does not mean that the water supply is safe forever, however. Because a water system can become contaminated quite suddenly, periodic testing is strongly recommended.

Accumulated evidence indicates that the bacteria causing typhoid, paratyphoid, cholera, and bacillary dysentery respond to their environments and to water treatment in a manner similar to that of coliform bacteria (Kehr and Butterfield, 1943).

Unfortunately, the test for coliform bacteria does not give absolute proof of the safety of a water supply. Certain viruses are also known to be more resistant than coliform bacteria to destruction by chlorine disinfection. From this, one can only conclude that it is possible for certain pathogenic organisms to be present in a water supply even though coliforms are not present. Tests exist for identifying these viruses, but in most cases, such tests are very complicated. Special tests are required for separate types of viruses, and costs for viral analyses are prohibitive for homeowners. In addition, coliform bacterial examination has historically served as a satisfactory measure of the microbiological safety of water.

Susceptibility of Water Sources to Bacteriological Contamination

When water seeps downward through the earth to the water table, particles in suspension, including microorganisms, may be removed. The extent of removal depends on the thickness and character of the overlying material. Clay, or hardpan, provides the most effective natural protection of ground water, although silt and sand also provide good filtration if they are fine-grained enough and occur in thick-enough layers. The bacterial quality of the water often improves during storage in the aquifer, because conditions are usually unfavorable for bacterial survival. Clarity alone does not guarantee that ground water is safe to drink; only laboratory testing can determine that.

In general, though, ground water found in unconsolidated formations such as sand, clay, or gravel and protected by similar materials from sources of pollution is more likely to be safe than water obtained from consolidated formations such as limestone, fractured rocks, or lava.

Where limited filtration is provided by overlying earth materials, water of better sanitary quality can sometimes be obtained by drilling deeper. It should be recognized, however, that there are areas where deeper drilling will not produce additional water because of the character of the local geology. Much unnecessary drilling has been done in the mistaken belief that more and better-quality water can *always* be obtained by drilling to deeper formations.

In areas without central sewer systems, human excrement is usually deposited in septic tanks, cesspools, or pit privies. Bacteria in the liquid effluents from such installations may enter shallow aquifers. In such areas, the threat of contamination may be reduced either by proper well construction or by location of the well farther from the source of contamination. Because the direction of ground water flow usually approximates that of the surface water flow, it is also always desirable to locate a well so that the normal movement of ground water flow carries the contaminant away from the well.

The fact that a spring is usually representative of ground water that is close to the earth's surface suggests that it has a greater potential for bacteriological contamination than water from a properly constructed well. Springs become even more susceptible to pollution when sources of contamination are located on higher adjacent land. For these reasons, some state and local health authorities recommend the use of a spring as a water supply only if the system is continually disinfected.

To safely select and use surface-water sources for individual water-supply systems requires consideration of additional factors not usually associated with ground water sources. When small streams, open ponds, lakes, or open reservoirs must be used as sources of water supply, the dangers of contamination and of the consequent spread of enteric diseases such as typhoid fever and dysentery are increased. As a rule, surface water should be used only when ground water sources either are not available or are inadequate. Remember that clear water is not always safe. The idea that running water purifies itself to drinking-water quality within a stated distance is false. In fact, the greater likelihood of physical and bacteriological contamination of surface water makes it generally unsafe for domestic use unless reliable treatment, including filtration and disinfection, is provided.

The treatment of surface water to ensure a constant, safe supply requires diligent attention to operation and maintenance by the owner of the system. Therefore, if ground water sources are limited, they should be developed for domestic purposes only. Surface-water sources can then provide for stock and poultry watering, gardening, fire fighting, and similar purposes. Treatment of surface water used for livestock is generally not considered essential. There is, however, a trend to provide stock and poultry with drinking water which is free from bacterial contamination and certain chemical elements.

CHEMICAL SAFETY OF WATER

Sources of Chemical Contaminants

Various chemical elements and compounds may be present in ground and surface water. The sources of these chemicals are associated with either natural processes or human activities.

Two important natural processes contributing chemicals to water are weathering and soil leaching. The factors controlling both the release of trace elements from rock and soil materials and their solution and stability in water are their solubility, pH, and adsorption characteristics. Decaying vegetation can also affect the concentration of chemicals in water. Many plants concentrate various elements selectively. As a result, some chemicals may be released during plant decay. Thus, rainwater infiltrating through the soil may pick up available chemicals that affect its quality. Likewise, runoff resulting from rainfall may transport chemicals to surface water (National Academy of Sciences, 1977).

Surface waters unaffected by human activities generally contain fewer chemicals and are softer than ground waters. However, hardness and dissolved solids in surface waters may vary considerably over the period of a year, or even, in many streams, from day to day. In a given stream, the hardness and concentration of dissolved solids generally varies inversely with the volume of water discharged, because the proportion of ground water, usually more concentrated than surface water, is much smaller during periods of high stream discharge than during periods of low stream discharge.

The average concentration of dissolved solids in the major rivers of the nation ranges from about 60 to 700 mg/L (ppm). Lower concentrations are found in streams in the Atlantic and Eastern Gulf Coasts, in the Pacific Northwest, and in the Great Lakes. Higher concentrations are found in the Western Gulf, the midcontinent, and the southwestern streams (ASTM STP 442, 1971).

Because of its long contact period with earth materials, ground water usually has higher dissolved solids and hardness than does surface water. However, ground water is also characterized by a more constant chemical composition and temperature.

The varying chemical quality characteristics of surface and ground waters are illustrated by the analyses given in Table 6-1. These analyses represent the raw water used for public supplies of eight cities. Analyses 1 through 4 are for surface waters and 5 through 8 are for ground waters.

Effects of Chemicals on Human Health

Some chemical elements found in drinking water are essential to human health. Copper, tin, iron, manganese, zinc, sodium, magnesium, calcium, molybdenum, cobalt, and vanadium are all necessary for human life functions. Very small amounts of chromium, arsenic, and selenium are also essential to human life. If some of these elements are ingested in excessive amounts, however, illness or death can result. Other elements, such as lead, mercury, and cadmium can gradually poison people by accumulating in human tissues until toxic levels are reached.

Scientific studies have demonstrated that metals such as lead or cadmium can displace the normal concentrations of zinc, manganese, and copper in the enzymes of the brain and other body tissues and thereby cause hyperactivity. An overabundance of copper can also replace zinc, manganese, and magnesium in certain tissues, leading to overstimulation, insomnia, elevated blood pressure, and restless nonproductive behavior.

Table 6-1. Analyses of Typical Surface and Ground Waters in the United States (*From U.S. Geological Survey Water-Supply Paper 1812*), *mg/L*

*Analysis number**	1	2	3	4	5	6	7	8
Date of collection	8/22/61	9/1/61	10/19/61	1961	8/9/61	9/15/61	8/23/61	7/25/61
Silica (SiO_2)	5.9	2.1	5.5	8.7	25	8.3	9.1	21
Iron (Fe)	0.00	0.17	0.40	—	0.03	0.72	0.10	0.01
Manganese (Mn)	0.00	0.14	0.00	—	—	0.01	0.06	0.00
Calcium (Ca)	8.5	28	37	84	65	12	57	26
Magnesium (Mg)	2.6	7.0	8.9	28	23	6.1	32	6.2
Sodium (Na)	3.6	4.1	17	92	14	7.5	3.4	86
Potassium (K)	1.5	0.9	5.1	4.0	1.8	0.7	1.7	1.9
Bicarbonate (HCO_3)	25	92	128	140	179	78	332	254
Carbonate (CO_3)	0	0	0	1	0	0	0	0
Sulfate (SO_4)	9.0	18	48	285	112	3.8	12	11
Chloride (Cl)	5.5	8.0	10	83	18	3.0	3.5	40
Fluoride (F)	0.1	0.0	0.4	0.4	0.8	0.4	0.1	0.7
Nitrate (NO_3)	4.7	0.5	3.2	1.4	0.2	1.2	0.2	0.0
Dissolved solids	59	129	222	657	410	87	284	318
Hardness as $CaCO_3$	32	99	129	323	256	55	274	90
Noncarbonate hardness as $CaCO_3$	11	24	24	206	110	0	2	0
Specific conductance (m at 25°)	91	213	324	1040	535	137	498	533
pH	6.4	7.6	7.7	8.4	7.9	6.8	7.6	7.2
Color	5	3	5	—	5	5	1	0

* Analysis numbers are identified as follows: 1. Baltimore, Md., North Branch Patapsco River (raw); 2. Detroit, Mich., Detroit River (raw); 3. St. Louis, Mo., Mississippi River (raw); 4. Los Angeles, Calif., Colorado River (raw); 5. Jacksonville, Fla., composite of several wells (raw); 6. Memphis, Tenn., Allen Well Field, composite of several wells 400 to 600 ft deep (raw); 7. Rockford, Ill., well 15, 1355 ft deep (raw); 8. Houston, Tex., Southwest Well Field, composite of 12 wells 490 to 2000 ft deep (raw).

Human health is affected by trace elements from many sources. Drinking water is one source, but it is not the most important. In general, the concentrations of chemical elements in foodstuffs exceed those found in drinking water. In some localities, abnormally large concentrations of certain chemicals are ingested from the atmosphere.

Still, many people are quite naturally concerned about the composition of the water they drink, and there is some divergence of opinion about what is ideal. For example, the effects of hard versus soft water have been hotly debated among various scientific, medical, and industrial groups. A similar controversy has arisen over fluoridation of water as a means of reducing dental cavities.

Chapter 5 provides a detailed description of the occurrence and health effects of inorganic chemicals that may be found in drinking-water supplies. Tables 6-2 and 6-3 list U.S. EPA's maximum contaminant levels for inorganic contaminants.

ORGANIC COMPOUNDS

Except in unusual circumstances, a sample of water is not tested for the presence of organic chemicals. However, industrial, agricultural, and domestic application of some of these chemicals may warrant analyzing for one or more specific organic compounds.

Over 300 organic compounds have been identified in drinking water. Some occur naturally, while others such as pesticides, are manufactured. The organic compounds that will be discussed or listed in this chapter have been selected for one or more of the following reasons:

1. Experimental evidence exists that on long-term ingestion these compounds are toxic to human or animal life.
2. Human ingestion over a lifetime may cause cancer or birth defects.
3. Its effect may be unknown at the present time, but the substance is commonly found in drinking-water supplies.
4. The compound is listed in the Safe Drinking Water Act or National Primary Drinking Water Regulations.

In general, concentrations of organic compounds in water are too low to produce acute toxicity, but the effects of long-term, continued ingestion might well represent a serious public-health problem.

Table 6-4 contains lists of organic compounds that are known or suspected to cause cancer in humans and animals. The numbers listed in the right-hand column estimate a person's cancer risk from a lifetime

Table 6-2. National Interim Primary Drinking Water Regulations: Maximum Contaminant Levels (MCLs) for Inorganic Contaminants Except Fluoride*

Contaminant	*MCL, mg/L*
Arsenic	0.05
Barium	1
Cadmium	0.010
Chromium	0.05
Lead	0.05
Mercury	0.002
Nitrate (as N)	10
Selenium	0.1
Silver	0.05

* The MCL for fluoride is determined by the annual average of the maximum daily air temperature for the location in which the community water system is situated.

of exposure to a particular compound. Debate continues concerning the validity of the scientific testing procedures and methods of arriving at specific concentration limits. However, these values represent the best current conclusions of present technology.

Table 6-5 lists organic pesticides and organic contaminants commonly found in drinking water. The ADI (acceptable daily intake) rep-

Table 6-3. National Interim Primary-Drinking-Water Regulations: Maximum Contaminant Level (MCL)* for Fluoride

Temperature		
°F	*°C*	*MCL, mg/L*
53.7 and below	12.0 and below	2.4
53.8 to 58.3	12.1 to 14.6	2.2
58.4 to 63.8	14.7 to 17.6	2.0
63.9 to 70.6	17.7 to 21.4	1.8
70.7 to 79.2	21.5 to 26.2	1.6
79.2 to 90.5	26.3 to 32.5	1.4

* Determined by the annual average of the maximum daily air temperature for the location in which the community water system is situated.

resents an empirically derived value reflecting current knowledge of the relative safety of each chemical.

These ADI values are tentative. They are meant not to represent a guaranteed safety level but rather to indicate an exposure level at which the chemical in question is not anticipated to produce an observable toxic response. The ADI values do not represent safe levels in drinking water, because they do not take into account what fraction of the potential contaminant intake may come from water. There is disagreement over the relationship of the part to the whole. For example, suggested no-observed-adverse-health-effects concentrations in water have been calculated under two different assumptions: (1) that 20 percent of total intake of a material is from water and 80 percent from other sources,

Table 6-4. Categories of Known or Suspected Organic Chemical Carcinogens Found in Drinking Water

Compound	*Highest observed concentrations in finished water* g/L	*Upper 95% confidence estimate of lifetime cancer risk for* g/L
Human carcinogen:		
Vinyl chloride	10	5.1 × 10-7
Suspected human carcinogens:		
Benzene	10	ID
Benzo (a) pyrene	D	ID
Animal carcinogens:		
Dieldrin	8	2.6 × 10-4
Kepone	ND	4.4 × 10-4
Heptachlor	D	4.0 × 10-5
Chlordane	0.1	1.8 × 10-5
DDT	D	1.2 × 10-5
Lindane (-BHC)	0.01	9.3 × 10-6
-BHC	D	6.5 × 10-6
-BHC	D	4.2 × 10-6
PCB (Aroclor 1260)	3	3.1 × 10-6
ETU	ND	2.2 × 10-6
Chloroform	366	3.7 × 10-7
Carbon tetrachloride	5	1.5 × 10-7
PCNB	ND	1.4 × 10-7
Trichloroethylene	0.5	1.3 × 10-7
Diphenylhydrazine	1	ID
Aldrin	D	ID
Suspected animal carcinogens:		
Bis (2-chloroethyl) ether	0.42	1.2 × 10-6
Endrin	0.08	ID
Heptachlor epoxide	D	ID

ID: Insufficient data to permit a statistical extrapolation of risk.
ND: Not detected.
D: Detected but not quantified.
Source: National Academy of Sciences, 1977.

Table 6-5. Organic Pesticides and Other Organic Contaminants in Drinking Water, Concentration, Toxicity, ADI, and Suggested No-Adverse-Effect Levels

Compound	Maximum observed concentrations in H_2O, g/L	Maximum dose producing no observed adverse effect mg/(kg)(day)	ADI++ mg/(kg)(day)	Suggested no-adverse effect level from H_2O, g/L Assumptions 1	2
2, 4-D	0.04	12.5	0.0125	87.5	4.4
2, 4, 5-T		10.0	0.1	700.0	35.0
Alachlor	2.9	100.0	0.1	700.0	35.0
Butachlor	0.06	10.0	0.01	70.0	3.5
Methoxychlor		10.0	0.1	700.0	35.0
Toxaphene		1.25	0.00125	8.75	0.44
Diazinon		0.02	0.002	14.0	0.07
Captan		50.0	0.05	350.0	17.5
HCB	6.0	1.0	0.001	7.0	0.35
Atrazine	5.0	21.5	0.0215	150.0	7.5
Propazine	Detected	46.4	0.0464	325.0	16.0
Simazine	Detected	215.0	0.215	1505.0	75.25
Di-*n*-butyl phtalate	5.0	110.0	0.11	770.0	38.5
Hexachlorophene	0.01	1.0	0.001	7.0	0.35
Methyl methacrylate	1.0	100.0	0.1	700.0	35.0
Pentachlorophenol	1.4	3.0	0.003	21.0	1.05
Styrene	1.0	133.0	0.133	931.0	46.5

Source: National Academy of Sciences, 1977.

and (2) that 1 percent of total intake is from water and 99 percent from other sources (NAS, 1977).

As shown in Table 6-1, whether derived from ground water or surface supplies, water in most cases meets U.S. EPA Primary Drinking Water Standards (Tables 6-2 and 6-3) for chemical constituents. When the chemical quality of water does reach standards unacceptable for human consumption, the contamination is usually due to the input of chemicals resulting from the activities of human beings.

HOW TO OBTAIN A SAMPLE FOR WATER-QUALITY ANALYSES

To obtain an adequate sample of water for either biological or mineral analysis, the objectives of sampling must first be established. Is a decision to be made between two or more potential sources for a water supply? Is bacteriological safety to be determined? Or is the interest primarily in discovering what water-treatment equipment is needed to ensure potable and palatable drinking water?

Of course, regardless of purpose, the sample taken must accurately represent the quality of the water source. One difficulty is that the concentrations of bacteria, dissolved gases, and minerals may vary with time, so it must be decided how best to preserve the constituents to be analyzed. This may call for refrigeration of the sample or the addition of chemicals to stabilize certain elements in the water. In some cases, it may be possible to transport the sample to a laboratory for analysis in a very short time. In others, due to water-quality constituents, it may be desirable to have equipment at hand for on-site analysis.

Knowing in advance what chemical or biological constituents are being sampled for may enable modification of sampling procedures or selection of more suitable sample methods. Knowing what is being sampled for is also a key to determining the urgency of sample preservation or how soon it must arrive at the laboratory for analysis.

Clearly, obtaining an adequate water sample involves more than turning on a water faucet and collecting some water in a jar. Human welfare may depend on determining the sanitary safety of a domestic water supply or discovering in it the presence of certain mineral constituents either injurious to health or aesthetically undesirable. Thus, one should, wherever possible, seek help in obtaining the sample.

The greatest problem will often be to decide who to contact. Various problems present various solutions. A sudden, unexplainable gastrointestinal illness should be reported to a local or county public health official, especially if there is reason to suspect that the domestic supply source is polluted. Should an unusual taste or odor suddenly appear in the water, or should the water become discolored, one could call a local distributor or sales representative for water-conditioning equipment. Some conditions, such as a change in water quality, may warrant a call to a local or county public-health official as well.

Appendix A at the back of the book lists the state agencies to which a homeowner may turn when a water-quality problem arises. These agencies will provide the homeowner with advice; in some cases they will even do the sampling necessary to ascertain the potential water-quality problem. However, most often they will suggest the appropriate local health official, university extension service, or private laboratory to contact for assistance. In localities where a public agency does the sampling and analysis, bacteriological analyses and possibly a limited chemical analysis will often be provided either at a nominal fee or at no charge to the homeowner. If an extensive chemical analysis is preferred, or if unusual chemicals in low concentrations are suspected of causing the complaint, then the homeowner will probably require the services of a private laboratory. However, 90 percent of the time, a county sanitarian or a county extension agent will be able to provide the needed assistance.

In any case, the essential first step in solving the problem is obtaining a reliable water sample. Wherever possible, a local, county, or state agency should run the analysis and send a technician to draw the sample. If a private laboratory is to do the analysis, it should send a technician to draw the sample. Homeowners who must obtain their own samples should request written instruction directly from the agency which will perform the analysis. These should include all relevant information on sampling techniques: where to sample and how to collect, package, preserve, and properly identify the sample. In some cases, the agency doing the analysis will provide a bottle or bottles in which to collect the sample. If the laboratory does not provide a sample container, or if the situation makes its use impractical, the final two sections of this chapter contain the needed instructions.

When the problem with water quality is not the result of a sudden change or is unrelated to health aspects, then a homeowner may find that a local water-conditioning salesperson may be able to provide the assistance needed. Local water-conditioning representatives are familiar with the water-quality problems typically encountered in a particular region. Frequently they can provide limited chemical analyses for water hardness, iron, pH, etc., and will usually be able to test the water for substances for which they sell or manufacture water-treatment equipment. Very often, a water-conditioning sales representative can provide quick assistance with most of the water-quality problems a homeowner may encounter. The names of local water-conditioning-equipment suppliers or sales representatives can normally be found in the Yellow Pages, or by contacting the Water Quality Association, Lombard, Illinois.

A water-well drilling contractor or a water-systems contractor may be able to run a few chemical tests on the water at no charge or for a nominal fee. If an old water system must be abandoned, then the drilling contractor or water-system contractor can provide cost estimates for materials and construction of a new system and also advise of any existing state or local construction codes that will affect the installation or design of a water system. Some homeowners do decide to construct their own water-supply systems, but in doing so they are taking considerable risks. The importance of a safe and reliable domestic water-supply system is obvious. Because of this vital importance, water-supply development should be left to experts familiar with codes and construction methods. The names of water-well contractors and water-system contractors can also be found in the Yellow Pages; additional assistance in locating contractors can be obtained from the National Water Well Association, Worthington, Ohio.

Private laboratories provide the most extensive water-analysis service. In some states they must be certified or licensed to run sanitary tests and to analyze the water for certain chemicals. Although well-equipped to analyze water samples, private laboratories are rarely able

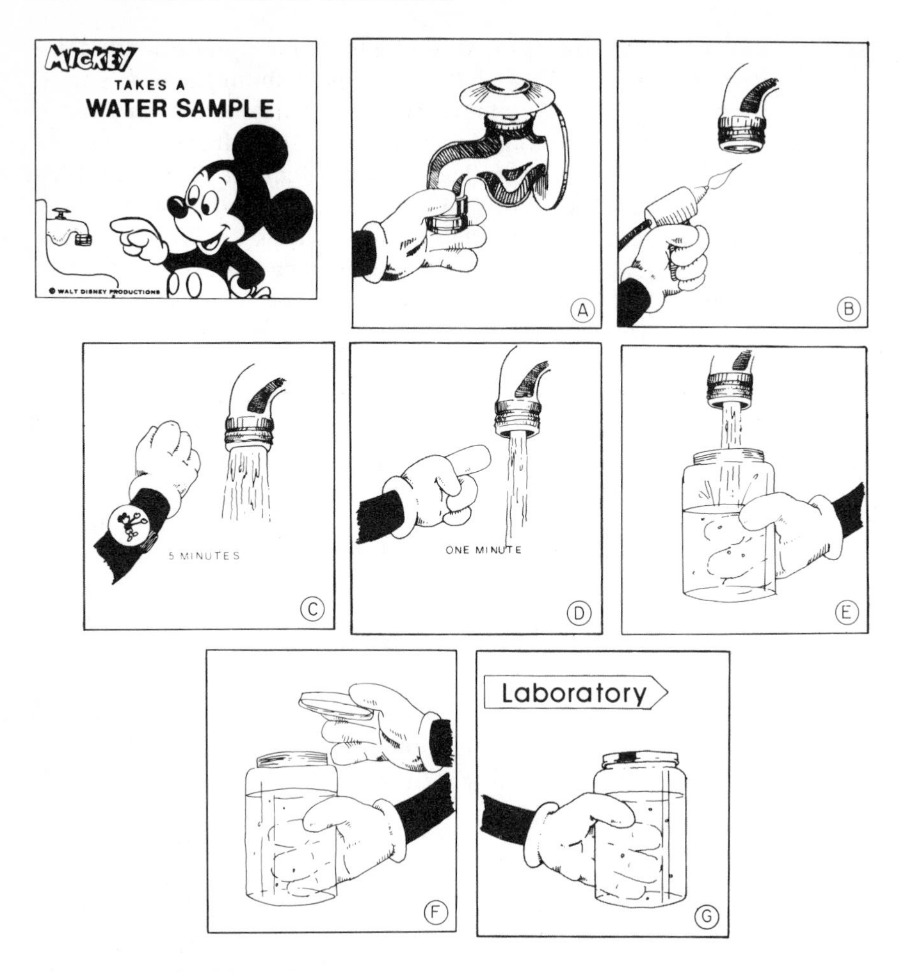

Figure 6-1. Checklist for collecting a water sample.

1. Select an indoor leak-free water faucet from which to take the sample.
2. Remove the faucet's aerator or strainer if one is present (A).
3. Flame the inside of the faucet with a propane or homemade torch to sterilize it. Do not wash, wipe, or touch faucet after sterilization (B).
4. Let water run at full flow for 5 min (C).
5. Close faucet to a stream of water the size of a pencil; let water flow for 1 min (D).
6. Fill bottle to ¾ capacity, while holding the bottle's cap in the other hand. Do not let anything touch the inside of the bottle or cap except the water sample (E).

to assist a homeowner in selecting a type of water-system construction or water treatment. Also, private laboratories may be considerably more expensive than public agencies or water-treatment-equipment suppliers. Many of the laboratories that perform water analyses on drinking-water supplies can be found in the Yellow Pages, or their addresses can be acquired from state or local health officials. The U.S. EPA also certifies laboratories that are equipped to provide services for water analysis in compliance with the Primary Drinking Water Regulations. However, a homeowner may select a laboratory not certified by the U.S. EPA.

HOW TO COLLECT, PRESERVE, PACKAGE, IDENTIFY, AND SHIP SAMPLE BOTTLES AND CONTAINERS FOR BACTERIOLOGICAL ANALYSIS

There is an old axiom that states, "The result of any test procedure can be no better than the sample on which it is performed."

The objective of sampling is to collect a portion of material small enough in volume—yet still accurately representing the material being sampled—to be conveniently transported to and handled in the laboratory. This implies, first, that the relative proportion or concentrations of all pertinent components must be the same in the sample as in the material being sampled; and second, that the sample be handled in such a way that no significant changes in composition occur before the tests are performed. (See Fig. 6-1.)

If a sterile bottle is not provided by the testing laboratory, then buy one from a pharmacy. If one is not available, select a pint or half-liter bottle, preferably one that is made of glass or other material resistant to the solvent action of ether and capable of withstanding heat during the sterilization process. A bottle with a screw cap is desirable, especially if the bottle is to be mailed. Either metal or plastic screw caps can be used, provided that no volatile compounds are produced during sterilization and that the caps are equipped with liners that do not produce toxic or bacteriostatic compounds when sterilized. Before the sample is taken, sterilize the bottle and cap by boiling in water for 15 min. After sterilization, take care to ensure that nothing except the water to be analyzed touches the inside of the bottle or the cap.

The outside of the faucet from which the sample is to be drawn should be inspected for leaks around the base of the packing gland

---

7. **Close bottle immediately after sample is taken (F).**
8. **Deliver sample immediately to laboratory or public agency, or store in the manner that they suggest. Samples not delivered within 30 hours must be retaken (G).**

while the faucet is running; if a leak is found, a different faucet should be selected. Only indoor, cold-water faucets should be used in sampling for bacteria.

Next, remove any aerator or strainer from the faucet. Sterilize the inside surfaces of the faucet by flaming with a propane torch. A homemade alcohol torch, consisting of a pliers-held cotton ball saturated with rubbing alcohol can also be used, but the flame of a propane torch is hotter. Do not wash or wipe the faucet after flaming until the sample is collected.

Open the faucet to full flow for 5 min or longer. Next, partially close the faucet so that a stream of water the thickness of a pencil is flowing. Maintain the flow this way for a minute prior to filling the bottle. When filling the bottle, hold it so that no water which comes in contact with your hands runs into the bottle. Also, do not set the cap down, but hold the bottle in one hand while holding the cap in the other. Fill the bottle to approximately three-quarters of its capacity. Close immediately after collecting the sample. Nothing should touch the inside of the bottle or cap except the water sample.

Immediately deliver the sample to the laboratory. If samples cannot be processed within 1 h, use preservation procedures recommended by the laboratory, or use iced coolers for storage. In no case should more than 30 h elapse between collection and analysis of the sample.

If the sample is drawn from a stream or other body of open water, the bottle should be plunged in at a point a short distance from the bank. Take care not to allow grass or other floating debris to enter the bottle.

Include the following information with the sample:

1. The owner's name and address
2. The type of water source (i.e., well, cistern, pond, etc.)
3. Where the sample was drawn (i.e., kitchen sink, bathroom)
4. Date and time of day the sample was drawn
5. Name and address of the person taking the sample
6. Procedure for sampling if different from instructions

TRACING THE SOURCE OF WATER CONTAMINATION

It is a basic rule of water sanitation to get to the source of a problem and eliminate it. If a well, for example, becomes contaminated, it is necessary to trace the contamination to its source and, if possible, remedy the situation. One tracing method is to empty a dye such as fluorescein into the suspected source of contamination. If the water supply becomes

Table 6-6. U.S. EPA Recommendations for Sample Storage and Preservation

Test	*Recommended volume, mL*	*Container*	*Preservative*	*Holding time*
Alkalinity	100	P, G†	Cool, 4°C	24 h
Arsenic	100	P, G	HNO_3 to pH 2	6 months
Chloride	50	P, G	None required	7 d
Color	50	G*	Cool, 4°C	24 h
Conductance, specific	100	P, G	Cool, 4°C	24 h
Fluoride	300	P*	Cool, 4°C	7 d
Hardness	100	P, G	Cool, 4°C, HNO to pH 2	7 d
Metals, total	100	P, G	HNO_3 to pH 2	6 months
Mercury, total	100	P, G	HNO_3 to pH 2	38 d (glass) 13 d (hard plastic)
Nitrogen, nitrate	100	P, G	Cool, 4°C, H_2SO_4 to pH 2	24 h
pH	25	P, G	Cool, 4°C, determine on site	6 h
Selenium	50		HNO_3 to pH 2	6 months
Solids, total dissolved	100	P, G	Cool, 4°C	7 d
Sulfate	50	P, G	Cool, 4°C	7 d
Sulfide	100*	P, G	2 ml of 2N zinc acetate	24 h
Turbidity	100	P, G	Cool, 4°C	7 d

* Hach recommendation.

† P: Plastic; G: Glass.

Source: Manual of Methods for Chemical Analysis of Water and Wastes, U.S. EPA, 1976, p. viii.

reddish or light green within a few hours, or even a few days, you have found the source of the contamination.

Because fluorescein is a strong coloring agent, it may be used whether the suspected contamination source is within a few hundred feet of your water supply or as far away as several miles. In a strong solution, it will be reddish in color. In a weak solution, it will be light green.

HOW TO COLLECT, PRESERVE, PACKAGE, IDENTIFY, AND SHIP SAMPLE BOTTLES AND CONTAINERS FOR CHEMICAL ANALYSIS

To obtain a sample for chemical analysis, follow the same basic procedures delineated in the previous section with the following important exceptions:

1. Contact the testing agency or laboratory to determine the size of the sample that would be required for a given chemical constituent.
2. When filling out the sample identification information, include a physical description of the water (i.e., color, turbidity, odor, and taste).

Based on physical characteristics of the water, such as taste, odor, or color, a state or local sanitarian can determine what chemical constituents should be determined in the sample.

Let as little time as possible elapse between the time the sample is collected and the time it will be analyzed. Depending on the nature of the analysis, when handling the sample special precautions may also be necessary to prevent natural alteration due to organic growth or loss of dissolved gases. Table 6-6 lists recommendations for sample size, holding time, and preservation techniques for analysis of different chemical constituents.

REFERENCES

ASTM STP 442, 1971: "Manual on Water," ASTM Spec. Tech. Publ., 3d ed.

Kehr, R. W., and C. T. Butterfield, 1943: "Notes on the Relation Between Coliforms and Enteric Pathogens," Public Health Report 59, 589.

National Academy of Sciences (NAS), 1977: *Drinking Water and Health*, pt. 1, chap. 5.

7

Treatment Techniques for the Removal of Taste, Odor, Color, and Turbidity

The importance of properly selecting and constructing a water-supply system has been described in preceding chapters, but neither selection nor construction practices can guarantee safe or aesthetically desirable water. Both surface and ground water frequently contain physical, biological, and chemical constituents that can affect the health of those using it for a water supply; lessen its aesthetic appeal; shorten the life of the water-supply system; and inconvenience the user.

The physical elements of water, such as taste, odor, color, and turbidity are described in this chapter, as are various treatment schemes for dealing with undesirable physical constituents of a water supply or system.

TASTE AND ODOR

Sensations of taste and odor are the result of chemical stimulation of the appropriate human nerve cells. Because of this, taste and odor are known as the "chemical senses." Taste and odor affect the quality of water in several ways, including reducing aesthetic desirability, tainting certain foods and beverages, and in some cases destroying the palatability of drinking water and foods cooked in water.

Oddly enough, the human nose may be the ultimate odor-sensing device. Most odors are detectable to humans at concentrations too low to permit their detection or definition by chemical or mechanical analysis. Therefore, laboratories must depend on human subjects for odor analysis. Tests have been developed to permit quality descriptions and approximate quantitative measurements of odor intensity.

A description of the laboratory tests for odor analysis would not be relevant to the needs of the average homeowner. For those interested, a detailed description of odor analysis is provided in APHA, 1975. (See References at end of chapter.) One should note that individuals may become acclimated to the odor of the water they drink, though houseguests may find it unpleasant.

When describing odor to a water-conditioning contractor it is wise to use familiar terms. For example, terms such as sweet, acid, and musty can be used to qualitatively define odor; terms such as weak, mild, and strong can be used to describe odor intensity.

Taste and odor differ in both the nature and location of the receptor nerve sites. Nerve sensors for odor are found high in the nasal cavity, while the nerve sites for taste are distributed over the tongue. In the case of a water sample, odor sensations are stimulated by vapors and do not require physical contact; taste sensations require contact of the taste buds with the water sample.

Tasting is a complex sensation: a combination of taste, odor, temperature, and feel that is called flavor. Taste tests usually have to deal with this complex combination. If a water sample contains no detectable odor and is presented at near-body temperature, the resulting sensation is predominantly true taste.

Like the test for odor, there are laboratory tests to evaluate the threshold of taste in a water sample and to determine the palatability of the water. However, homeowners are generally concerned only with the taste their water may have and how they, their family, and their guests will react to it as a drink, mixed with another beverage, or used in the preparation of food. As a homeowner, the following subjective rating scale can be used to evaluate the taste and odor of water (from APHA, 1975).

1. I would be very happy to accept this water as my everyday drinking water.
2. I would be happy to accept this water as my everyday drinking water.
3. I am sure that I could accept this water as my everyday drinking water.
4. I could accept this water as my everyday drinking water.
5. Maybe I could accept this water as my everyday drinking water.
6. I don't think I could accept this water as my everyday drinking water.
7. I could not accept this water as my everyday drinking water.
8. I could never drink this water.

9. I can't stand this water in my mouth and I could never drink it.

Causes of Odor and Taste in Water

Undesirable tastes and odors in water are normally a result of a combination of circumstances. The materials that contribute to these observed tastes and odors come from many sources. Table 7-1 lists examples of the taste or odor characteristics that certain inorganic chemicals may impart to water. Although some inorganic compounds can, in concentrations of a few parts per million, impart taste and odor to water, the main sources of odor- and taste-bearing substances are organic materials. These materials, under certain conditions, can cause persistent difficulties when present in only trace amounts of a few parts per billion (Tables 7-2 and 7-3). Wastes from chemical plants, refineries, and individual onsite waste-disposal systems (cesspools, septic tanks, etc.) have a great impact on water's palatability. However, biological constituents such as bacteria and algae are the most frequent causes of bad taste and odor. (Sigworth, 1957.)

The metabolic activities of some algae produce an intense, unpleasant odor. Actinomycetes are another source of odor in water, imparting an earthy or musty odor. Nematodes produce an oily, gummy substance that also has an earthy or musty odor.

Problems of taste and odor in water do not usually present any particular health hazard, but people are naturally concerned that the water they have to drink be at least palatable and, if possible, pleasant-tasting.

TREATMENT TECHNIQUES FOR THE ELIMINATION OF TASTE AND ODOR PROBLEMS

Most taste and odor problems are dealt with by eliminating the substance that causes the problem. Therefore, once the cause of the prob-

Table 7-1. Odor and Taste Characteristics of Common Inorganic Chemical Constituents Found in Water

Chemical	*Taste or odor characteristics in water*
Iron	Bitter taste
Manganese	Bitter taste
Sulfate	Bitter taste
Hydrogen sulfide	Rotten-egg odor
Sodium chloride	Salty taste
Bicarbonates	Flat, soda taste
High TDS content	Salty taste

Table 7-2. Concentration of Some Chemicals Causing Taste and Odor

Substance	*Concentration detectable, ppb*
Formaldehyde	50 000
Picolines	500–1000
Phenolics	250–4000
Xylenes	300–1000
Refinery hydrocarbons	25–50
Petrochemical waste	13
Chlorinated phenolics	1–100

Source: F. M. Middleton et al., "Taste and Odor Research Tools for Water Utilities," *J. AWWA* vol. 50, p. 21, 1958.

lem is identified, the proper treatment system can usually be selected. Treatment techniques for taste and odor can be divided into three major categories: (1) filtration; (2) demineralization; and (3) disinfection.

Activated-Carbon Filtration

Activated-carbon filtration systems are widely used to remove unpleasant tastes and disagreeable odors from water. These systems are successfully used to remove tastes and odors caused by industrial wastes, pesticides, decaying organic matter, dissolved gases, and residual chlorine and the by-products of chlorination.

Activated carbon "adsorbs" the taste- and odor-producing im-

Table 7-3. Threshold Odor Concentration (TOC) of Some Individual Compounds

Compound	*TOC, ppb*
Naphthalene	6.8
Tetralin	18
2-Methyl-5-ethylpyridine	19
Styrene	37
Acetophenone	65
Ethylbenzene	140
Bis(2-chloroisopropyl)ether	200
2-Ethylhexanol	360
Di-isobutyl carbinol	1300
Phenylmethyl carbinol	1450

Source: A. A. Rosen et al., "Relationship of River Water Odor to Specific Organic Contaminants," *J. Water Pollut. Control Fed.*, vol. 35, p. 777, 1963.

purities in water. Adsorption is defined as the adhesion of a gas, vapor, or dissolved material on the surface of a solid. Adsorption can be differentiated from absorption in the following manner. A particle of activated carbon has an extremely large surface area (400 to 500 m^2/g) owing to its structure of pores similar to those found in a sponge. But a sponge will absorb water containing a taste or odor, and when the water is squeezed out of the sponge, the taste and odor will still be present in the water. In adsorption with activated carbon, the water is brought into contact with the carbon particles and the taste and odor constituents are retained on the carbon, resulting in a taste- and odor-free water.

The adsorptive capacity of activated carbon is directly related to the surface area of the material that comes in contact with the water. Thus manufacturers of activated carbon attempt to maximize pore space.

There are two basic types of activated-carbon-filter systems: (1) the cartridge filter; and (2) the activated-carbon bed filter.

The cartridge-type filters are very popular today and are manufactured by a wide variety of companies. Small cartridge units are designed to fit on the drinking water tap to remove undesirable tastes and odor-causing substances from water used for drinking and cooling (Fig. 7-1). The entire cartridge is periodically replaced. Replacement is recommended when either tastes or odors reappear or the water pressure from the tap starts to decrease noticeably. The decrease in pressure is caused by particles building up on the surface of the filter.

Larger cartridge filters are designed so that the filter can be removed and cleaned by washing off the accumulated dirt particles. After several cleanings, when the carbon material has adsorbed all the taste and odors it can hold, the filter is replaced. The larger cartridge filters are engineered for more extensive use and are usually mounted under a sink.

The bed-type activated-carbon filter (Fig. 7-2) is designed to treat most of the water being used in a home. It is usually placed beyond the chlorinator and/or water softener. The bed-type filter must be backwashed periodically to remove suspended dirt from the filter bed. Backwash frequency and instructions are usually supplied by the manufacturer, and proper maintenance is necessary to ensure satisfactory performance.

Gradually, the carbon bed's ability to eliminate tastes and odors is reduced because the pore spaces become saturated, and it must then be replaced. For an average household, the activated carbon will last from about 1 to 3 yr depending on the intensity of the taste and odor being removed, as well as the quantity of water which it is conditioning.

A more detailed description of the equipment operation and maintenance procedures for activated-carbon-filtration systems is provided in Chap. 11.

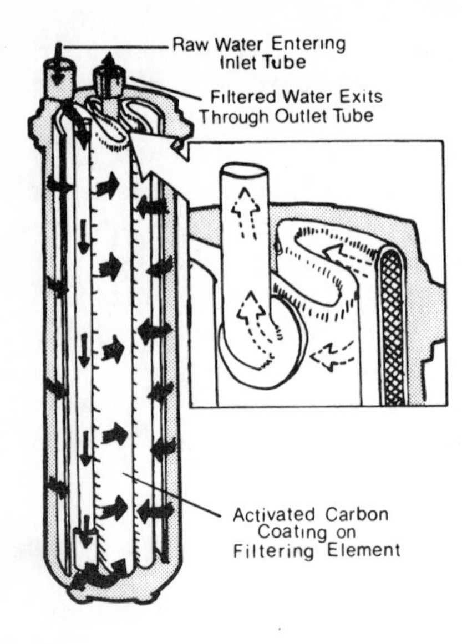

Figure 7-1. One model of cartridge-type activated-carbon filter. Water is filtered as it passes from the outside of the filter element into the inner area. The inset shows how filtered water inside the element is collected and discharged through a special outlet. (*From American Assoc. for Materials, 1973.* "Planning for an Individual Water System.")

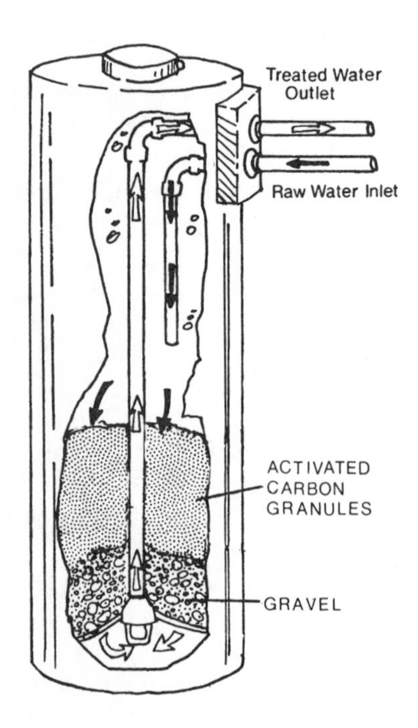

Figure 7-2. Activated-carbon–bed-type filter in the service mode of operation. Water flows through the control valve, down through the activated-carbon bed. The activated carbon will adsorb chlorine and other taste- and odor-producing organic matter until its capacity is exhausted. The filtered water then flows up through the distributor system, back through the valve, and to the water-supply system. Black arrow indicates unfiltered water. White arrow indicates filtered water. (*F. E. Myers and Bros., Co.*)

Chlorination

In addition to filtration of water to remove taste- and odor-causing substances, strong oxidants can be used to change taste- and odor-causing substances to innocuous forms. Chlorination, used primarily for disinfection purposes, is probably the most popular of the oxidation methods. Because chlorine also controls the growth of algae and microorganisms, it has the ability to reduce the quantity of these taste- and odor-bearing substances in a water system.

A detailed description of use and performance capability of chlorination is provided in Chap. 8.

Potassium Permanganate

Potassium permanganate, steadily gaining in water-treatment use since about 1960, acts effectively as an oxidizing agent to destroy tastes and odors. It has been used successfully to destroy taste and odor resulting from the presence of dissolved hydrogen sulfide gas, and metallic ions such as iron or manganese. Because of its ability to oxidize soluble iron and manganese to insoluble oxides, some method of filtration should follow the potassium-permanganate treatment to remove chemical precipitates. Normal dosage of potassium permanganate for domestic water treatment ranges between 1 and 5 ppm, varying with water-quality conditions.

Ozonation

Ozone is a powerful oxidizing agent that has been proven to remove effectively many types of tastes and odors. Ozone's ability to remove tastes and odors is inherent in various stages of the ozonation process. First, the process includes aeration, which strips out gases and volatile chemicals. Suspended organic particles may be removed by neutralizing the electric charges that keep them separated, so that they coagulate into larger particles that can be removed by filtration. Iron and manganese are oxidized into insoluble hydrated oxides and hydrogen sulfide is oxidized to insoluble, black, amorphous elemental sulfur. Ozone then destroys odor- and taste-producing bacteria and algae.

Because of the chemical precipitates and organic particles produced by ozonation, some type of filtration following the process is desirable to remove particulates.

The ozonation system involves passing dry, clean air through a special form of high-voltage electric discharge. The mixture of air leaving the ozone generator contains about 1% ozone (O_3), which is passed through the water to be treated.

Ozonation is a relatively new process as related to domestic water treatment. Although the technology has been with us for about a century, it has generally been applied to large water-treatment plants. On a domestic level there are still some developmental problems in some systems. Also, the initial cost and the cost of operation of the system are higher than the costs of other methods available for the removal of taste and odor.

Chapter 8 contains a more detailed discussion of the application of ozonation for disinfection of water.

Aeration

Aeration is the process of bringing about the intimate contact between air and a liquid such as water. Certain types of tastes and odors are removed or reduced by aeration.

Aeration is most practical for treating surface-water sources. Water is allowed to fall over a spillway in a turbulent stream or it is distributed in multiple streams or droplets through a series of perforated plates. A simple aeration unit may consist of a vertical holding tank which permits dissolved gases to seep out of the water into the atmosphere.

Although the aeration of water may be accomplished in an open system, adequate precautions should be exercised to eliminate possible external contamination of the water. Whenever possible, a totally enclosed system should be provided.

Aeration may be used to reduce odors caused by the presence of hydrogen sulfide and algae. However, care must be taken in removal of hydrogen sulfide by aeration because it is flammable. Aeration is also effective in increasing the oxygen content of water deficient in dissolved oxygen. The flat taste of cistern water may be improved by adding oxygen.

Aeration may not be feasible in areas which experience freezing temperatures.

Copper Sulfate

The most frequent source of taste and odor in an individual water-supply system is algae. These minute plants produce certain biological by-products which cause tastes and odors in the water. Under some circumstances, chlorination of the water will accentuate the tastes and odors.

When the source of water is a pond or a lake, copper sulfate treatment can be used to kill algae or control their growth.

The dosage of the copper sulfate varies with the particular species of organism involved. A dose of 0.3 mg/L, however, will generally control all but a few of the growths likely to cause trouble in drinking water. For certain species of fish, particularly those of the trout family, this dosage may be poisonous. The approximate doses of copper sulfate (in milligrams per liter) which should not be exceeded to avoid killing various kinds of fish, are given in Table 7-4.

In small reservoirs or ponds, the required dose of copper sulfate can be dissolved in water and introduced by a sprinkling can. In large ponds or reservoirs, copper sulfate may be tied in a clean gunnysack and dragged through the water from a boat in lanes 10 to 20 ft (3 to 6 m) apart until the copper sulfate is completely dissolved.

The span of time over which a treatment will be effective varies, depending upon sunshine, reseeding, and local conditions. Generally, several treatments per season are required. Some municipalities treat open distributing reservoirs as often as twice a month in order to avoid unexpected blooms of algae and the accompanying taste and odor problems. In such a case, control is easier and more effective if the treatment is done before the algae bloom or reach their maximum growth and development.

Commercial algicides for use in swimming pools are widely available but are not intended for use in water designated for human, livestock, or poultry consumption. They should not be used for domestic water-supply treatment.

Maintenance of a continuous and adequate chlorine residual will effectively control the growth of algae in controlled storage facilities.

Additional information is provided in Chap. 8.

Table 7-4. Approximate Maximum Tolerance Limits of Various Fish to Copper Sulfate

	Copper sulfate dosage	
Kind of fish	*mg/L*	*lb/gal* $\times 10^6$
Trout	0.15	1.2
Carp	0.30	2.5
Suckers	0.30	2.5
Catfish	0.40	3.5
Pickerel	0.40	3.5
Goldfish	0.50	4.0
Perch	0.70	6.0
Sunfish	1.20	10.0
Black bass	2.00	17.0

Diatom control. Diatoms are another form of algae, recognizable by their regular boxlike or tubular walls of silica and brownish-green color. When found in colonies, they form a variety of geometric patterns—disks, cylinders, plates, etc.

Diatoms commonly produce geranium and violet aromas in low concentrations, fishy and moldy odors in high concentrations.

As is the case with other kinds of algae, diatoms generally yield to treatment with copper sulfate.

Because algae require sunlight for growth, problems of taste and odor resulting from their existence can be eliminated by covering the water source to prohibit their exposure to light.

Removal of Petrochemicals

The problem of petrochemicals has been placed in this section of the text because the presence of hydrocarbons is usually detected by taste or odor.

Hundreds of reports of water-well contamination by oil or other petroleum products are investigated each year by public-health agencies. The degree of contamination is usually related to the size of the pollution source. Large areas of an aquifer have been made unpotable because of oil that has entered from an improperly cemented or abandoned oil-production well. Very localized areas of contamination have resulted from oil that has been spilled during transportation and storage. In general, the larger the area of the aquifer affected, the more difficult it is to remedy the situation. In many cases, one can only wait for nature to take its course and, over a period of many years, purify the aquifer. However, in many situations small spills occurring through leakage of a fuel-storage tank or transportation accident can be corrected.

Often the source of the problem may be oil from an oil-lubricated pump, either used in the well or used to pump water from a cistern or a surface source. In such cases, the problem can easily be remedied by a few adjustments followed by pumping the source of water until all the oil is removed. However, cases do exist where farmers or homeowners have stored fuel on their property and either the storage tank has leaked or spillage has occurred during a delivery. Eventually these petrochemicals can be washed into a surface source of water or migrate to the water table. When ground water is the source of water, the effects of a petrochemical spill may not become apparent for many months or even years due to the slow movement of petroleum products through the ground.

If the problem is minor, the petrochemicals may be removed by installing one or more charcoal filters right after the storage tanks. If the water is to be chlorinated, it should be done after the removal of the hydrocarbons with carbon filters so as not to produce undesirable trihalomethanes.

Imbiber beads, produced by Dow Chemical, Inc., can be used for treating persistent problems caused by larger concentrations of hydrocarbons in water-supply systems. These beads can be obtained through the manufacturer or local distributors and come in packets or "pillows" of varying sizes. The size selected depends on how and where the pillow is to be used. Pillows 4 in by 4 in (10 cm square) are most convenient for use in domestic water-supply systems. They float on water and remove free oil, oil in solution, and a number of other petrochemicals. A cord is tied to each pillow, then the pillows are dropped into the water-storage tank, or into the well if ground water is being used. The cords permit easy removal of the pillows when they become saturated with petrochemicals.

Imbiber beads will either remove all the oil and petrochemicals or lower their concentration to a point where charcoal filtration will work. After imbiber beads have been used, the hot-water tank should be drained and flushed.

Removal of Iron and Manganese

Iron and manganese may impart a metallic taste to water and are therefore undesirable. Most of the methods for removing taste and odor which rely on oxidants (i.e., potassium permanganate, ozone, etc.) produce hydrated oxides of these metals that are insoluble in water.

Another method of removing iron or manganese is through a process of ion exchange. The ion-exchange process is described in detail in Chap. 9.

TURBIDITY AND COLOR

Turbidity is a haziness in water caused by the presence of insoluble suspended particles. It is also defined as a lack of clarity or brilliance in a sample of water.

Turbidity is usually due to suspended inorganic or organic substances. These suspended particles range in size from fine colloidal to coarse grains of sand that will remain in suspension only as long as the water is agitated.

Generally, turbidity is more common in surface water than ground water, because soil and sedimentary rocks filter water moving through the ground. In surface water, turbidity starts to increase in the late spring and peaks in later summer and fall as a result of the addition of organic substances to the water and to the "turnover" of lakes and ponds.

Turbidity is measured by determining the amount of light that is scattered by particles in a sample of water. Too high a reading indicates a potential danger to the water-supply system. A maximum limit of 1 turbidity unit is recommended because certain types of turbidity-causing solids, such as organic matter, can interfere with microbiological determinations, or can prevent maintenance of an effective disinfectant residual throughout the distribution system.

Pure water is colorless or light blue. Water that is discolored contains foreign substances, generally organic compounds derived mainly from soil humus which, in turn, is produced from the decay of plant or animal matter.

Natural metallic ions, such as iron and manganese, are sometimes the cause of color in water. Water can also be discolored from industrial wastes or mine runoff.

The color intensity of water is determined by comparing a sample of water with distilled water containing differing concentrations of platinum-cobalt solutions. Highly colored water is objectionable because it may stain household fixtures and clothing, as well as reduce aesthetic appeal.

TREATMENT TECHNIQUES FOR THE REMOVAL OF TURBIDITY AND COLOR

Turbidity and color in water can be reduced or removed by any one technique or a combination of a number of techniques, used either in combination or separately: coagulation, sedimentation, filtration, and oxidation.

Coagulation

Coagulation is a process which causes fine suspended particles to collect into large particles that can settle to the bottom of a pond, lake, cistern, or holding tank so that clear water can be removed from near the surface. Coagulation is initiated by adding to the water, a chemical, such as alum (hydrated aluminum sulfate), which induces suspended particles to combine physically and form a floc.

Factors which affect coagulation and flocculation rates are pH and temperature of the water, size of the particles, rate at which the coagulant and water are mixed, and the ultimate quantity of coagulant used.

Sanskrit medical lore and Egyptian inscriptions describe the earliest use of coagulation for water treatment, dating back as far as 2000 B.C. Coagulation was accomplished by using a variety of mineral and vegetable substances, principally the seed of *Strychnos potatorum*. Other substances found to be effective coagulants include an amazing variety of materials, among which were certain nuts, beans, and pounded barley.

Alum, the coagulant most widely used today, was known to the early Egyptians. First mention of its use as a coagulant for water conditioning was made by the Roman author, Pliny (A.D. 77), who described both lime and alum as useful for rendering bitter water potable. By the fifteenth century alum was being produced on a commercial scale for water-treatment applications. During the second half of the nineteenth century, the use of alum coagulation for treatment of water supplies grew in popularity throughout the United States. In addition to alum, iron salts such as ferric chloride and ferric sulfate have been used as coagulants.

The dosage of a coagulant should be determined by a trained technician, since the quantity and type of coagulant used will depend on water turbidity, temperature, and pH. To determine the amount of turbidity in a water supply, a sample must be analyzed. Suppliers of filters, or the local health department, may be able to perform this analysis.

If alum, which is acidic, is the coagulant selected, it must be applied before the water is chlorinated.

When the pH of the water ranges between 6.8 and 7.5, alum is the most suitable coagulant. Ferric sulfate can be used when the pH range of water is between 5.5 and 8.8 (AWWA, 1971, p. 88).

Some generalizations can be made with regard to the use of coagulants: (1) A minimum coagulant must be added for clay turbidity to provide an enmeshing mass floc. (2) Some additional coagulant is generally required with increase in turbidity, but the dosage of coagulant will not increase linearly with increase in turbidity. (3) Paradoxically, where turbidities are very high, smaller dosages of coagulant will often suffice because of the high collision probabilities. (4) A broad distribution of clay-particle sizes is much easier to coagulate than a suspension containing a single size of particle or a narrow range of particle sizes. (AWWA, 1971, p. 87.)

Coagulation-flocculation treatment is generally applied to surface-water sources where time and space are available for flocculation to take place without the need of an additional settling tank. If a settling tank is used, it will require periodic cleaning to remove mud and floc that accumulate on the bottom. With a well-designed system, cleaning should have to be done as infrequently as twice a year.

Sedimentation

Sedimentation is the process by which floc or suspended matter is separated from water by precipitation and deposition.

Like coagulation-flocculation, sedimentation has a long history in water treatment. In ancient Egypt, the turbid floodwaters of the Nile were diverted to stilling basins and allowed to settle before distribution for domestic use. In the ruins of Carthage, the remains of large rainwater storage and settling reservoirs, erected for the public, have been found.

Sedimentation can be accomplished in a quiescent pond or properly constructed tank or basin. Basically, the process depends on the effect of gravity on particles suspended in a liquid of lesser density.

Under the influence of gravity, particles having a density greater than water will settle. But at least a 24-h detention time must be allowed for a significant reduction in suspended matter. The tank inlet should be arranged so that the incoming water containing suspended matter is distributed uniformly across the entire width as the water flows to the outlet located at the opposite end. Baffles can be used to reduce high local velocities and short-circuiting of water. To facilitate cleaning and repairing, an installation should be designed with two separate sections, each of which may be used independently.

Settling tanks require periodic cleaning to remove sediment or sludge. The frequency of the cleaning is determined by the size of the settling tank, the turbidity of the water, and the rate at which the water is used.

Filtration

Filtration is the process of removing suspended matter from water as it passes through beds of porous material. In the process, suspended silt, clay, colloids, and many types of microorganisms are removed. The degree of removal depends on the character and size of the filter media and the size and quantity of the suspended solids.

A wide variety of filtration systems exists, including filters for inlets of surface sources of water; rapid-sand filters; upward-flow filters; diatomaceous earth filters; and cartridge filters. Pads, ceramic cylinders, paper, porous stone, and slow-sand filters are not recommended for filtering turbid water. They are not satisfactory because of their low filtering capacity, the ease with which water can channel, their tendency to crack, their inability to filter out fine material, or the ease with which they may clog.

How well a filter functions will depend on factors such as filter area; the quality of water to be filtered; the porosity of the filtering material; required flow rates; and design capacity of the filter between servicing and filter construction.

Filter area is an important aspect of any filter system because it affects the capacity and the flow rate. The larger the filter area, the greater the potential flow rate. Therefore, to size a filtration system properly, flow rates must be known in advance of system construction.

The size of the particles that will do the actual filtering must be considered, because the smaller the particles of the filtering material, the finer the particles of suspended matter that can be removed.

Filter construction is very important. If the water can find a channel or by-pass around or through the filter material, it will remain unfiltered.

Filtered Inlets for Surface Water

To prevent dead organic matter or suspended particles from entering the water system through its source at the surface, the intake pipe should have a screen or filter on it. The filtered inlet should neither rest on the surface of the water nor remain at the bottom of the source, for turbidity is usually greatest at these localities. It is best to attach a float to the inlet pipe so that it remains 12 to 18 in (30 to 45 cm) below the surface of the water, even when the water level fluctuates. (Fig. 7-3)

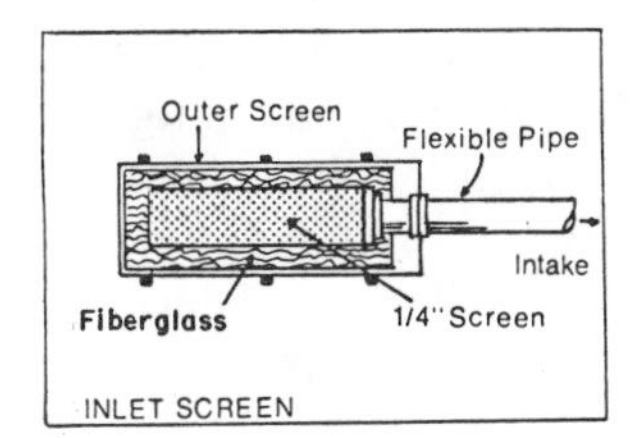

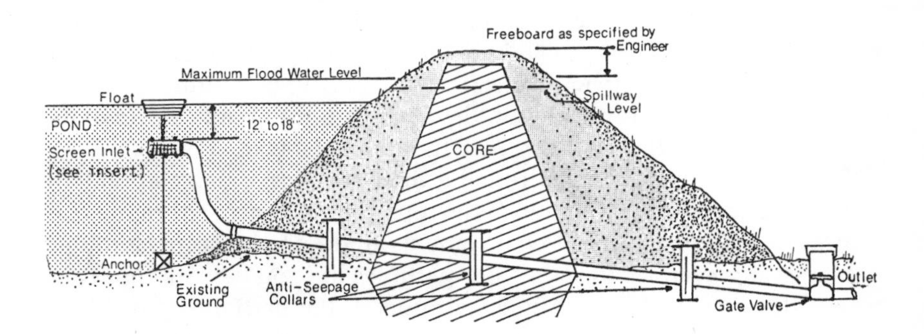

Figure 7-3. Schematic diagram of a surface-water system where a screened, floating inlet is utilized.

Conventional Pressure Filters

Conventional filters consist of a single layer of sand about 24 in (60 cm) deep. The conventional sand filter has particles about one-fiftieth of an inch in diameter (0.5 mm). After backwashing the smallest particles are at the top and the larger, heavier particles are at the bottom of the bed. Water enters at the top of the filter, flows down through the bed and out the bottom. Particles of dirt are trapped at the top of the unit, and a "deck" of turbidity forms there. When the pressure loss becomes excessive, the filter is backwashed to clean up the turbidity from the filter bed.

Conventional filters are usually designed to remove suspended material which would cause turbidity. A quartz sand is normally used as the filter media. However, several different types of granular filter material can be selected to remove other undesirable constituents in the water (i.e., granular activated carbon for some taste and odor removal; manganese greensand for removal of iron or manganese; or limestone chips for neutralizing acid water).

The filters are cleaned by backwashing water through them in a direction opposite to that used during the filtration process itself. The backwash serves to clean the suspended solids from the filter media and carry them to waste. Systems that require manual backwashing should be cleaned once a week, whereas an automatic sand filter is normally backwashed every day. There must be enough pressure in the system to suitably backwash the filter. In fact units should be sized on the basis of their backwash capability. If the system lacks suitable pressure for backwashing on large filter, then two or more smaller filters with individual backwash capability can be used.

Multimedia Filters

A multimedia filter uses three or four layers of various materials stacked to a total bed height of 30 to 36 in (75 to 90 cm) (Fig. 7-4). The top layer consists of large, lightweight particles of plastic or bituminous coal. Rather than act as a trap material, their purpose is to keep the top of the filter bed open so that dirt can penetrate into the bed.

The second layer of a multimedia filter bed is normally anthracite coal with particles half the size of the top layer. Because anthracite coal is heavier than plastic, the coal remains underneath the plastic after backwashing even though the coal particles are smaller. As the water passes through the layer of filtering mineral, the particles of turbidity come in contact with the surface of each granule. Coagulation and flocculation can then occur within the filter bed. This process allows tiny particles to build up into larger particles which are then trapped in this portion of the filter bed.

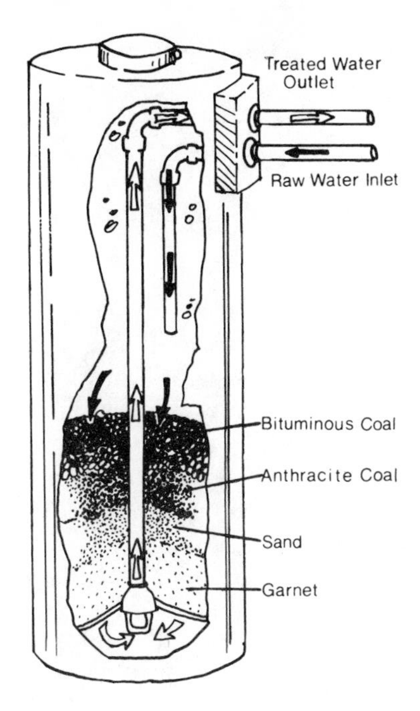

Figure 7-4. Multimedia Filter.

The third layer of filtering media is sand, as in a conventional sand filter. The particle size of the sand granules is about half the size of the coal above it. The turbidity does not form a deck at the top of the sand layer because the water is pushing the particles through the sand bed. Additional contact with the granules allows finer turbidity to grow into large particles and become trapped within the sand layer.

Finally, the water reaches the bottom layer, usually garnet, of the multimedia filter. This semiprecious mineral is extremely fine and weighs 30 percent more than sand. Its minute particles trap a much finer turbidity than is possible with conventional sand filters. Consequently, the water is filtered to a higher degree of clarity.

Because turbidity is trapped throughout the entire bed rather than merely in the very top layer, the multimedia filter can operate several times longer than a conventional sand filter before backwashing is required.

Conventional sand filters must be operated at a slow flow rate to trap the particles at the top layer. For example, a 12-in-diameter (30-cm) household filter should be operated at 1.5 to 3.5 gal/min (5.5 to 13.25 L/min). This ensures good water clarity without turbidity breakthrough. By comparison, a 12-in multimedia filter can easily be operated at 7.5

gal/min (30 L/min). Both filters require about 10 gal/min (40 L/min) for proper backwashing.

Multimedia filters are more expensive than conventional sand filters of the same size because of the costly minerals used in the tank. This investment can sometimes be offset by selecting a smaller-diameter tank which will perform as well as a large-diameter sand filter.

Precoat Filters

The precoat water filter is entirely different from the conventional type of filter. Even its size is substantially smaller. It consists of a pressure vessel (container); a porous septum material; and a compound of filter media. When water is initially run through the filter, the powdered filter medium mixes with the water and forms an even coat or deposit on the septum, thereby forming the precoat filter cake. As the water flows through this cake, its solid impurities (iron, dirt, cloudiness, etc.) are caught on the cake's small pores. When the filter medium is an activated-carbon-based material, the adsorption of tastes and odors (including those of chlorine) takes place as the water passes through the thickness of the filter cake.

Diatomite Filters (Diatomaceous Earth Filters). Diatoms are a type of tiny marine algae that multiply prodigiously throughout the world's oceans. These tiny organisms produce glasslike skeletal structures that are deposited on the sea floor. Diatomaceous earth is the skeletal remains of diatoms. A cubic inch of diatomaceous earth may contain the remains of 50 million diatoms. The shell or skeleton of this organism consists of lacy halves, some resembling snowflakes in design, which fit together like a pillbox. The large surface area provided by the shell allows the organism to float. After death, however, the same shell becomes an excellent material for a filter medium. The diatomite filtering element usually consists of a porous surface called a *septum*. The septum can consist of wire cloth, plastic fiber cloth, or any of several materials that will let water pass through readily. A coating of diatomaceous filter materials is then applied to the septum to provide filtering action. The diatomaceous filter material retains the suspended solids in the liquid and protects the filter element from becoming clogged. The septum itself acts as a check filter, since it must be constructed with exposed pore sizes which are small enough to allow the diatom shells to bridge across the openings.

Maintenance consists of adding new diatomaceous earth when the filtering action decreases. Eventually, the time comes when adding additional diatomaceous earth will have little effect on improving filtration. At this stage, the entire filter must be removed and cleaned and new diatomaceous filter material must be added to the septum. Under normal use, replacement should not be required more than once every 2 months.

Diatomaceous earth filters have been used successfully to remove turbidity from ponds and rivers, remove algae, reduce iron and manganese content, and, to a lesser extent, remove some chlorine-resistant pathogenic organisms.

Cartridge-Type Filters

Cartridge-type filters have grown in popularity and are now produced by a wide variety of manufacturers; they are readily available through water-system contractors, water-conditioning specialists, and department stores. They are designed to fit on a service line just before a single tap (i.e., kitchen faucet) and are generally used to filter only water used for drinking and cooking.

Most cartridge filters are capable of removing sediment and suspended particles that cause turbidity. If the cartridge contains activated carbon, it may also be useful in removing tastes and odors and some by-products of chlorination. However, labels should be carefully read to evaluate the capability of the cartridge filter being considered for purchase.

The cartridge filter owes its popularity to its simplicity of installation and maintenance. The entire unit consists of a pressure container and a cartridge element. Generally, only a periodic change of the cartridge element or the entire cartridge is required. Other advantages include maintenance of water quality: as the cartridge becomes older only the flow rate changes. Also, the filter element is corrugated to allow maximum filter area in the volume of the pressure container. This large amount of filter area allows some manufacturers to use a very fine filter material which may result in extremely fine filtration.

Some of these filters provide excellent fine filtration; others provide only marginal coarse filtration. Care is necessary in selecting the proper filters.

Separators

Separators are generally used to remove sand or silt from well water. However, it should be mentioned that if a well is properly constructed, the water should be sand-free.

While separators vary in design, they are all forms of a hydrocyclone. Although it has been known for years (the first United States hydrocyclone patent was issued in 1891), such equipment did not come into common use until the early 1950s, and then it was employed basically in only the mineral-recovery and mining fields.

Hydrocyclone devices have a tangential connection from the liquid source and a vortex finder contained in a cylindrical or conical separation chamber. The velocity of the flow accelerates solids entering the separation chamber, thus exerting extremely high centrifugal forces on the particles. The solids are forced to the outer walls of the separation chamber and move in a spiral path along the wall to the apex of the cone-shaped separator or to the collection chamber of a cylindrical device (Fig. 7-5). Meanwhile, the clarified or separated product, now free of the heavier particles, moves in toward the vortex, or void, in the center of the separation chamber and moves upward through the vortex finder to the discharge. (Anonymous, 1973).

For domestic use, most people select an in-line separator, as shown in Fig. 7-6. These units are capable of removing up to 98 percent by weight of all suspended solids in water. They can remove particles as small as 0.0029 in (74 μm).

Over the years many submersible well pumps have been ruined by sand entering the impellers. Now separators are designed to remove particulates before they enter the pump. These units are made to fit submersible pumps with diameters as small as 4 in (10 cm), or to fit casings with minimum diameters of 6 in (15 cm).

Algicides

Turbidity can be due to the presence of algae in the water. Some methods of treatment for controlling algae have been described earlier in this

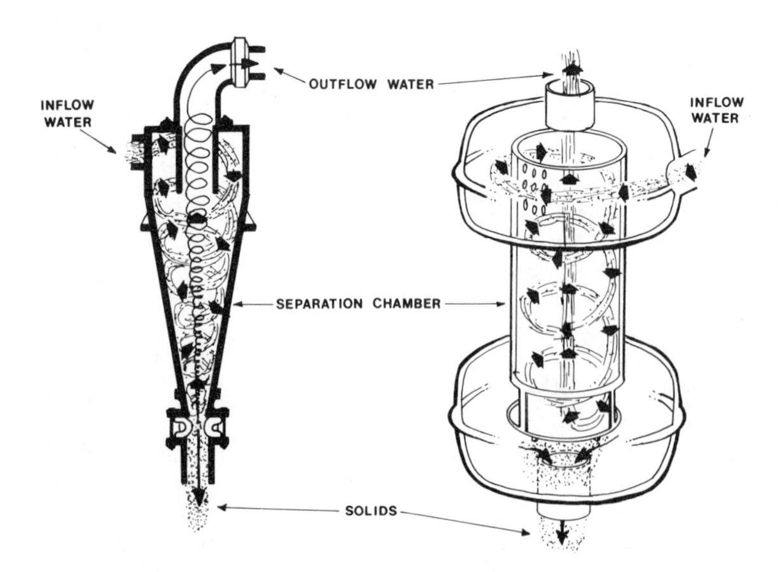

Figure 7-5. Hydrocyclone separator.

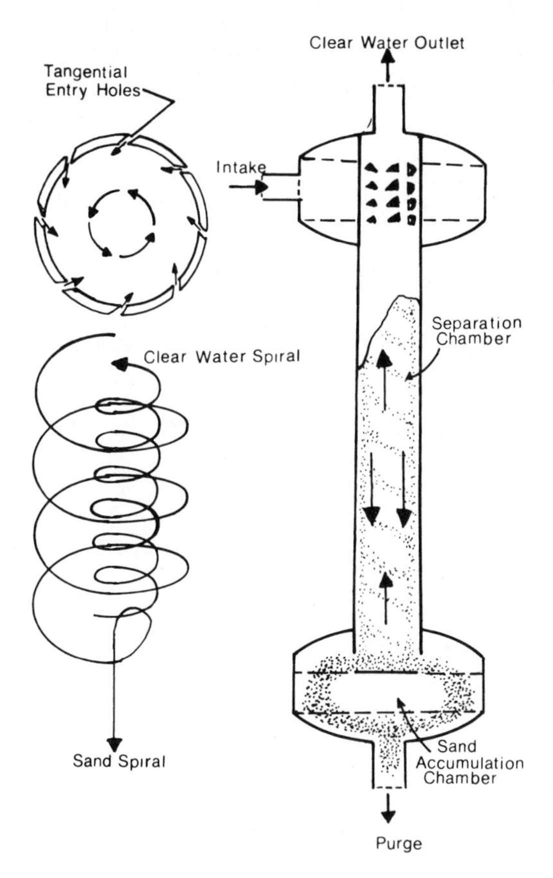

Figure 7-6. In-line separator designed for domestic or agricultural use.

chapter. An in-depth description of the types and uses of algicides is presented in Chap. 8.

CONSUMER NOTICE

Not all the equipment on the market to treat water-quality problems may meet sales representative's or manufacturer's claims. The consumer should use the same discretion and caution that would be used in buying any health-related product or household appliance. Shop around and check with a local health agency if performance claims seem questionable.

REFERENCES

American Association for Vocational and Instructional Materials, 1973: *Planning for an Individual Water System.*

American Public Health Association (APHA), 1975: *Standard Methods for the Examination of Water and Wastewater*, 14th ed., APH-AWWA-UPCF.

American Water Works Association (AWWA), 1971: *Water Quality and Treatment*, 3d ed., McGraw-Hill, New York.

Anonymous, 1973: *The Separator Story, Ground Water Age*, vol. 8, no. 1, pp. 24–28, September.

Myers, F. E., and Brother Co., 1968: *Myers Water Conditioning Technical Manual*, Ashland, Ohio.

Rosen, A. A., et al., 1963: "Relationship of River Water Odor to Specific Organic Contaminants," *J. Water Pollut. Control Fed.*, vol. 35.

Sigworth, E. A., 1957: "Control of Taste and Odor in Water Supplies," *J. AWWA*, vol. 49.

8

Treatment Techniques for the Removal of Biological Contaminants

INTRODUCTION

All natural waters, regardless of source, are likely to contain some microbial organisms. A few cause disease; some impart taste, odor, or turbidity to the water; others are beneficial; and the rest are of little interest.

Surface-water supplies are highly susceptible to biological contamination from exposure to the environment (especially to runoff, and atmospheric particulate matter) and should receive continuous disinfection.

Water which percolates through soil to become ground water usually has comparatively fewer micro-organisms because of the natural tendency of soil and subsurface materials to filter water. In addition, oxygen is rapidly depleted in the subsurface environment, the result of which is deactivation or death of some micro-organisms. However, even ground water can easily become contaminated by domestic sewage, feedlots, surface runoff, etc. Certain subsurface formations may not adequately remove biological constituents. In highly fractured rocks, in cavernous limestone, or where relatively shallow ground water sources are tapped, natural filtration may not adequately remove biological constituents and polluted ground water may migrate many miles.

Thus, no water supply is immune to dangerous contamination, and purity should never be assumed. For this reason, most public water utilities have elected to chlorinate their water. This has resulted in a substantial reduction in severe outbreaks of waterborne diseases in areas served by public water systems. However, reports of waterborne disease outbreaks have been increasing since 1950 owing to

1. An improved ability to diagnose waterborne diseases
2. The failure of water-treatment systems
3. The failure to treat individual or small community water-supply systems for biological contaminants.

The average annual number of waterborne disease outbreaks in the United States since 1938 is shown in Fig. 8-1. (Center for Disease Control, 1976.)

The most important outbreaks of waterborne infectious diseases that occurred from 1971 to 1974 are listed in Table 8-1. The causative agent was determined in only 53 percent of the 99 disease outbreaks that involved 16 950 cases (Craun, McCabe, and Hughes, 1976). The remainder were categorized as "acute gastrointestinal illness of unknown etiology."

ORGANISMS WHICH CAUSE WATER PROBLEMS

Bacteria

Most water bacteria are harmless, and many are actually beneficial. They consume organic detritus, thereby reducing the water's chemical

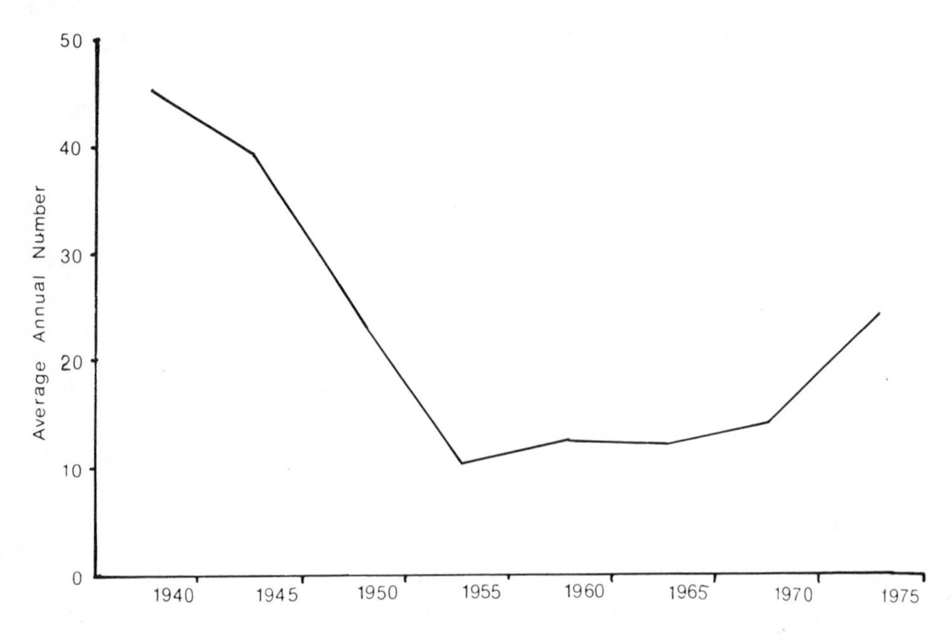

Figure 8-1. Average annual number of waterborne disease outbreaks, 1938–1975.

Table 8-1. Etiology of Waterborne Disease Outbreaks and Cases, 1971–1974

Disease	*Outbreaks*	*Cases*
Gastroenteritis	46	7992
Giardiasis	12	5127
Shigellosis	13	2747
Chemical poisoning	9	497
Hepatitis-A	13	351
Typhoid	4	222
Salmonellosis	2	37
Total	99	16 950

Note: Because so many cases of gastroenteritis are never officially diagnosed, it is difficult to determine accurately the frequency of these waterborne diseases.

oxygen demand; prolong the useful life of filters; and destroy some foul tastes, odors, and colors. In addition, some produce by-products which kill or inhibit the growth of some pathogens. Since pathogens are usually less numerous and less hardy in water than are other bacteria, their survival can be decreased by ordinary competition for food with more vigorous organisms.

There are several groups of bacteria which are harmless from a disease point of view but undesirable because they are nuisances. One group is the actinomycetes, which are filamentous like molds, only smaller. These are the most common producers of musty and earthy tastes and odors in water. Another group specializes in transforming sulfur-containing materials into hydrogen sulfide, the familiar rotten-egg smell. A similar group includes the iron and manganese bacteria. When they become established in plumbing carrying iron-bearing water, they produce acids that cause severe blocking or corrosion of pipes and thus supply their growing need for more iron to utilize. The taste, coloration, and staining of the water will increase until they are killed, they clog the pipes entirely, or they corrode completely through.

There is a large subgroup of innocuous, "neutral" bacteria known as "opportunistic pathogens." These organisms can cause infections, contribute to already established diseases, lengthen or complicate recovery, or otherwise cause unwanted effects in individuals already debilitated by some other illness. They become pathogens when presented with an unusual opportunity.

In general, pathogenic bacteria are best suited to the environment inside the body and therefore do not fare very well when excreted into water. They usually do not multiply, for example, and most of them are more easily killed by disinfecting chemicals than the other bacteria naturally present in water. There are some notable exceptions, though, which are quite hardy and persistent; and rapid pathogen die-off should never be taken for granted.

Some pathogens form spores, which are highly resistant to disinfectants in their dormant forms. Although none of the sporeformers are direct human waterborne pathogens, three are notable: *Bacillus anthracis* from water causes anthrax in animals, which can then be transmitted to human beings. *Clostridium tetani* could conceivably cause tetanus ("lockjaw") if spores were introduced into deep wounds during bathing. *Clostridium botulinum* produces botulism toxin, the most powerful poison known. The spores will not infect human beings directly, but foods prepared with spore-laden water may develop the toxin if stored sealed, and they can then be deadly if consumed without cooking. Recently, infant botulism is being suspected as one of several causes of the sudden infant death syndrome (SIDS), or "crib death" of children under the age of 6 months. Botulism spores are so resistant to chemical treatment that they must be either physically removed by filtration at the submicrometer level or subjected to pressure-cooker temperatures for an hour.

Cholera is the most serious waterborne disease; death often occurs within hours of onset if medical treatment is not provided promptly. It is transmitted by personal contact, by carelessly prepared food, or by polluted water. The causative organism, *Vibrio cholerae* (formerly *V. comma*), dies off rapidly in average surface waters but persists for weeks in either very clean or very turbid water. Turbidity also protects it from disinfection procedures.

Salmonellosis is the waterborne disease most easily traced to its source by laboratory detective work. There are several hundred species and variants of the genus *Salmonella* known to attack human beings. Their effects range in severity from typhoid fever, caused by *S. typhosa* (the only species which attacks human beings exclusively), to the common acute intestinal upsets formerly known as "ptomaine poisoning." The source can be direct or indirect fecal contamination from practically any warm-blooded animal.

Shigellosis is the most common waterborne cause of acute diarrhea in the United States. Most epidemics result from personal contact or carelessly prepared food, but waterborne outbreaks are not rare. As with *Salmonella* there are many *Shigella* types, the most serious being *S. dysenteriae*, the cause of bacillary dysentery.

Although usually considered harmless, *Escherichia coli*, the predominant intestinal inhabitant of humans and lower animals, occasionally causes diarrhea in susceptible individuals. More common in infants, "enteropathogenic," "enterotoxic," or "enteroinvasive" *E. coli* can be serious. The dozen or so infective strains are often harbored in, and spread by, persons exhibiting no symptoms. A similar but not identical malady, known as "travelers' diarrhea," or "Montezuma's revenge," is caused by regional variants of *E. coli.* These are harmless to natives of the locality but troublesome to visitors.

Yersinia enterocolitica was first described in 1965. By 1970 it ranked as a major cause of human gastroenteritis in Scandinavia. By 1974 it ranked number one in France and Hungary. In 1976 it was prominent in the United States and Canadian Rockies. It is being identified more often every year, probably because researchers are learning how to isolate it, but also because it may possibly be adapting and spreading.

Yersinia enterolitica thrives in waters colder than 10°C. Under such conditions the water usually appears quite pristine. It is rather chlorine resistant (similar to poliovirus), but this may be the effect of reduced chlorine activity at low temperatures. Six biotypes have been identified—four in North America and two in Europe. Persons affected by *Y. enterolitica* eventually recover, but unless treated with antibiotics, they may remain disease carriers.

Leptospirosis is comparatively uncommon, but its occurrence is worldwide. Caused by a group of a hundred or more types of the genus *Leptospira,* the disease is variously known as Weil's disease, febrile jaundice, or swamp fever. The organisms can be excreted by most domestic and wild animals, and streams carry them many miles. People most often become infected while swimming, but bathing and drinking water are also possible sources.

Tularemia is readily spread to the human population via drinking water contaminated by wild animals in rural areas, although it is most commonly acquired by direct contact with animals in rural areas or with their insect parasites during hunting season. The infectious agent, *Francisella tularensis* (formerly, *Pasteurella tularensis*), is unusually virulent (infective), as few as five organisms being sufficient to produce the disease in laboratory animals; they can survive in water for weeks or months, depending on temperature.

Tuberculosis is commonly thought to be an air- or aerosol-spread disease of the lungs, but it has been spread by swimming in or drinking contaminated water. *Mycobacterium tuberculosis* and its close relatives cause not only the classic lung disease but also granulomatous infections of the skin, eyes, and ears. The organisms are chlorine resistant, being able to survive rather well in the presence of up to 2.5 mg/L. In untreated water and sewage, they persist with virulence for perhaps half a year.

Algae

Algae are simple organisms, often existing as single, microscopic cells, which manufacture their own food from sunlight and nutrients. Some, however, do form large structures, and often cause problems because of the sheer bulk of their enormous numbers. Several species produce chemical substances which impart musty or fishy tastes and odors to water. Others produce slime, which interferes with water-treatment processes. During periods of rapid growth ("blooms"), they cause water to be turbid and colored, and at these times clogging of water filters can become very serious. Some marine types are toxic, but no waterborne diseases have been directly attributed to freshwater algae.

Fungi and Molds

These filamentous organisms frequently form dense, slimy mats which clog filters and other water-treatment equipment. They can also produce musty tastes and odors, as well as color and turbidity.

Viruses

Viruses are the smallest of infectious agents, some being as small as single protein molecules. These very simple particles are not actually considered to be living organisms. They exist exclusively as parasites on the cellular level, invading individual cells of more complex creatures and forcing their host's cellular machinery to help them multiply. In general, they are both more virulent (a single particle of some viruses can cause disease) and more resistant to disinfection than bacteria. Fortunately, they are present in water in far fewer numbers than bacterial pathogens, by a factor of 100 to 1000. Medical treatment of infected individuals is limited to complicated manipulations of the body's immunity system, since the "wonder drugs" such as penicillin, aureomycin, etc., are useless against them. For viruses, the best defense is prevention in the forms of good sanitation practices and a continuing, vigorous program of inoculation.

Theoretically, any virus which is excreted by human beings or lower animals can be spread by water and cause disease, although only polio and hepatitis have been specifically traced and identified. Laboratory methods for isolating, counting, and identifying viruses found in water are very difficult, expensive, and unreliable, and are therefore not undertaken regularly. Waterborne intestinal upsets, infections, common-cold-type illnesses, etc., which have no other obvious causes, are often attributed to viruses, even when direct evidence is lacking.

ANIMAL ORGANISMS PRESENT IN WATER

Protozoa

Even more than is true of bacteria, the vast majority of protozoa present in water are beneficial. They consume quantities of organic waste materials, and they also help to keep both algae and bacteria within the bounds of a balanced ecology. They do not produce taste, odor, color, or turbidity, and their growth is self-limiting since it is dependent on oxygen. There are only two major pathogenic protozoa, *Entamoeba histolytica* and *Giardia lamblia*.

Entamoeba histolytica is the cause of amoebic dysentery, a more serious and persistent disease than bacillary dysentery. Unfortunately, it is also more common, and its primary route of transmission is contaminated drinking water. It is infectious only in the cyst stage, a highly resistant, dormant form similar to bacterial spores. Amoebic cysts share with hepatitis virus the distinction of being the most chlorine-resistant pathogens. However, the cysts are rather large (8 to 12 μm in size) and are effectively removed by fine filtration in the micrometer range. *Entamoeba coli* is also pathogenic to human beings, but it is less serious and less common.

Giardia lamblia is a protozoan which causes a recurring diarrhea when ingested in drinking water. Like *Entamoeba*, *Giardia* is also chlorine resistant in the cyst stage, but it is easily removed by fine filtration. Individual organisms are generally 12 to 15 μm in diameter.

There are also other intestinal amebas: *Endolimax* and *Iodoameba;* an always fatal brain ameba, *Naegleria;* and a ciliate, *Balantidium,* which parasitize human beings from water; but they are much less common.

Parasitic Worms

There are three groups of parasitic worms which trouble human beings: roundworms (nematodes), unsegmented flatworms or flukes (trematodes), and segmented flatworms or tapeworms (cestodes). Roundworms are transmitted mainly through consumption of their eggs, although some may enter the body directly through the skin. Flukes are just the opposite, being most often acquired while swimming or wading. They almost invariably require an intermediate host in their life cycle, usually a snail. Tapeworms are always acquired by consuming the eggs. The infective forms of all parasitic worms are best removed by fine filtration, since chemical disinfectants may be only marginally effective.

METHODS OF DISINFECTION

Depending on the water source, disinfection may be a simple or complex matter. Ground water from deep, drilled wells may not need disinfection if regular bacteriological tests show that it is safe. Shallower wells or dug wells almost certainly will require disinfection, and surface waters probably would benefit from fine filtration as well.

There are a number of ways of purifying water. In evaluating the methods of treatment available, the following points regarding water disinfectants should be considered:

1. The disinfectant should be effective on many types of pathogens and on whatever numbers may be present in water.
2. The length of retention time required should be sufficient for the disinfectant to kill all pathogens in the water.
3. The disinfectant should function properly regardless of water-flow fluctuations.
4. The temperature and pH range in which the disinfectant will be required to function must be adequate.
5. The disinfectant must not make the water either toxic or unpalatable.
6. The disinfectant should be safe and easy to handle.
7. The concentration of the disinfectant in the water should be easy to monitor.
8. The disinfectant should provide residual protection against possible recontamination.
9. The disinfectant should be readily available at a reasonable cost.

Because no single type of disinfectant meets all these criteria, a certain amount of compromise is necessary in making a selection.

The type of equipment used to dispense the disinfectant during the water-treatment process is as important as the disinfectant. Desirable characteristics include the following:

1. The system should be automatic, requiring only a minimal amount of maintenance for proper operation.
2. The system should be fail-safe, so that no one can unwittingly use or consume contaminated water.
3. The treatment should provide a residual effect to ensure that any organisms escaping initial treatment or introduced into the system beyond the treatment device will be destroyed.
4. The treatment process should affect all water entering the home, eliminating the complications and dangers that exist

if one tap yields contaminated water while another produces safe water.

Because no system can meet all these criteria, the system selected, like the disinfecting agent itself, will strike a compromise among the various criteria.

Filtration

There are two kinds of filtration: particulate removal, which is for the most part a physical sieving or straining process, and adsorption, which is mainly chemical in nature. Adsorption is the attraction and bonding of certain dissolved substances, and sometimes of extremely tiny particles, onto the surfaces of large particles, using a force which acts something like magnetism. Adsorption filtration is commonly used to remove taste- and odor-causing organic substances, chlorine, pesticides, other toxic organics, and even asbestos fibers and viruses. Depending on the design, construction, and type of medium used, a filter may be able to remove any one, or possibly many types of contaminants from water, to varying degrees.

In the treatment of surface waters which may contain large, resistant parasites such as protozoan cysts and worm eggs, fine filtration (fine-particulate removal) is a necessary companion to chemical disinfection. Although filters capable of removing 99.9 percent of bacteria from water may be purchased, they are not intended for this purpose. No filter is perfect; some bacterial and viral pathogens can always be expected to pass through, so one should never depend on filtration to do the entire job of disinfection. The types and ranges of fine-particulate filtration methods are the following:

1. Precoated filters using finely powdered filter media such as activated carbon and diatomaceous earth.
2. Resin-bonded fiber filters.
3. Cast ceramic filters.
4. Polymeric membrane filters—the above four filters are in the 2.0- to 0.2-μm range.
5. Ultrafilters, which remove large molecules and large viruses.
6. Reverse osmosis and electroosmosis, which remove viruses of all sizes, small molecules, and even atoms.

NOTE: Filters designated as "granular filters," "bed filters," or "depth filters," that do not have a specific (absolute) micrometer rating of 3 μm or less are not acceptable for the removal of protozoan cysts and larger organisms.

Chlorination

Chlorine is a disinfecting agent used extensively to treat water for municipal and individual supplies. Its popularity is owed to the fact that both the disinfectant and the treatment system meet nearly all the desirable criteria described in the preceding section.

Chlorine is commonly available in three forms: solid, liquid, and gas. Municipalities and community water systems frequently use chlorine gas (Cl_2) for water purification. In its gaseous state, however, chlorine is not safe to handle, and the equipment required to deal with it is too expensive for use in treating individual water-supply systems. Liquid sodium hypochlorite is commonly used in domestic chlorination systems. Sold in grocery stores as household bleach, this product consists of a 5.25% solution of sodium hypochlorite, which is equivalent to 5% available chlorine. Other available sodium hypochlorite solutions range in strength from 3 to 15% available chlorine by weight. These solutions can be diluted with potable water to produce the desired solution strength. Chlorine solutions should be stored in a cool, dark place if they are to maintain their designated strength because light produces a photochemical reaction that reduces their potency.

Chlorine is available in dry form as calcium hypochlorite. It can be obtained commercially as either a soluble powder or as tablets marketed under such trade names as B-K Powder, H.T.H., Perchloron, and Pittchlor. These compounds are classed as high-test hypochlorites because they contain 65 to 75% available chlorine by weight. Packed in cans or drums, they are stable and therefore will not deteriorate if properly stored and handled.

When hypochlorite powders are used, fresh chlorine solutions should frequently be prepared, because the strength of the chlorine solutions deteriorates gradually after preparation. The containers or vessels used for preparation, storage, or distribution of chlorine solutions should resist corrosion and the effects of light. Suitable materials include glass, plastic, crockery, and rubber-lined metal containers. When hypochlorite powders are used, it will be necessary to decant the solution periodically to remove sediment from the chlorine container.

Hypochlorite solutions can be used either full strength as prepared or diluted to the solution strength required by the feeding equipment and the rate of water flow. In preparing these solutions, one must consider the chlorine content of the concentrated solution. For example, if 5 gal (19 L) of 2% solution are to be prepared with a high-test calcium hypochlorite powder or tablet containing 70% available chlorine, the high-test hypochlorite needed would weigh 1.2 lb (0.54 kg).

Pounds of compound required

$$= \frac{\%\ \text{strength of solution} \times \text{gallons solution required} \times 8.3}{\%\ \text{available chlorine in compound}}$$

$$= \frac{2 \times 5 \times 8.3}{70}$$

$$= 1.2\ \text{lb}\ (0.54\ \text{kg})$$

Expressed in another way, 1.2 lb of high-test hypochlorite with 70% available chlorine when added to 5 gal (19 L) of water will produce a 2% chlorine solution.

Chlorination Equipment. Chlorine disinfectant solutions are put into water with small chlorinator units, of which there are three types. The pump types are called positive-displacement, or PD, feeders. They are electrically powered and operate by alternately expanding and compressing a rubber form (like a plunger) or by moving a piston (like a syringe). The eductor-type chlorinators use the natural vacuum created by the flow of water in a pipe to draw the chlorine solution from the disinfectant reservoir. There are tablet or granule-type feeders which allow solid disinfectants to contact the flowing water to be treated. As material dissolves, more is forced down into the dissolving chamber by gravity.

POSITIVE-DISPLACEMENT FEEDERS. The most common kind of positive-displacement hypochlorinator uses a piston or diaphragm pump to inject the solution. This type of equipment, which is adjustable during operation, can be designed to operate at reliable and accurate feed rates. The starting and stopping of electrically powered hypochlorinators can be synchronized with those of the pumping unit. Although hypochlorinators can be used with any water system, they are especially desirable in systems where water pressure is low or fluctuating.

Figure 8-2 illustrates the mechanism used in one type of positive-displacement feeder, including (1) the electric motor; (2) two check valves; (3) a pumping chamber; (4) a diaphragm, and (5) drive-block assembly consisting of a cam and drive shaft that connects the motor with the diaphragm.

A special type of injection check valve is required to ensure that the proper amount of chlorine enters the mainstream of water flow. Figure 8-3 illustrates one type of injection system. When such systems are not used, the alternatives too often result in inadequate chlorination, rapid clogging, or corrosion of the pipe around the injection point.

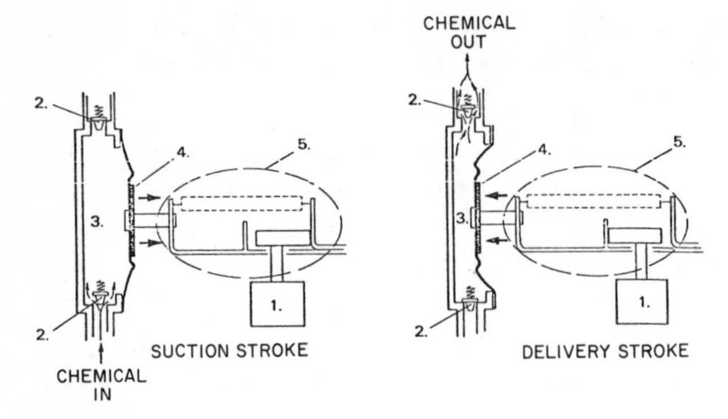

Figure 8-2. A common type of positive-displacement feeder. As the diaphragm (4) is pulled back, the volume of the chamber (3) is increased. A partial vacuum is formed, pulling the liquid through the lower check valve (2) and into the chamber. The diaphragm then moves forward, the lower check valve closes, and the liquid is forced through the upper check valve (2) to the point of injection.

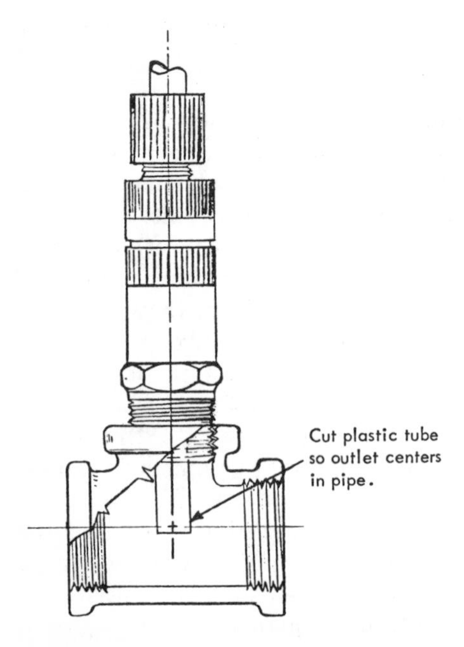

Figure 8-3. A common type of injection valve.

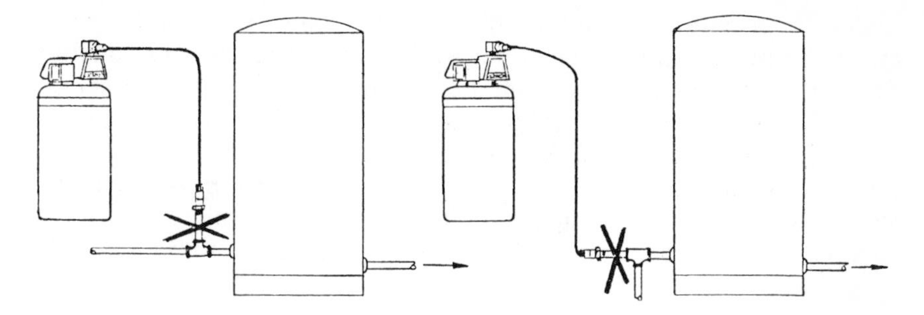

Figure 8-4. Improper installation techniques. *Wrong!*—Do *not* use pipe stub in conjunction with tee for insertion of injection valve. Do *not* insert pipe stub into tee. This keeps injection point out of water line. The full-strength solution will either corrode through the stub or clog it solid with scale.

Figure 8-4 shows some common installation errors. A pipe stub should not be used with a tee for insertion of the injection valve nor should it fit into the tee, because the injection point would project above the waterline. The full-strength chlorine solution would then either corrode the stub or clog it with scale.

Figure 8-5 illustrates the proper installation procedure. The chlorine solution is injected into the water supply before it reaches the pressure tank to ensure that the chlorine mixes completely with the water and that the contact time is sufficient for chlorine disinfection to occur. Pipe corrosion can result if dilution does not occur at the injection point. This problem is easily prevented by installing the injection fitting so that the tip is in the center of the line carrying the water being treated.

EDUCTOR FEEDERS. The two most common eductor feeders are aspirator feeders and suction feeders. The aspirator feeder operates on a simple

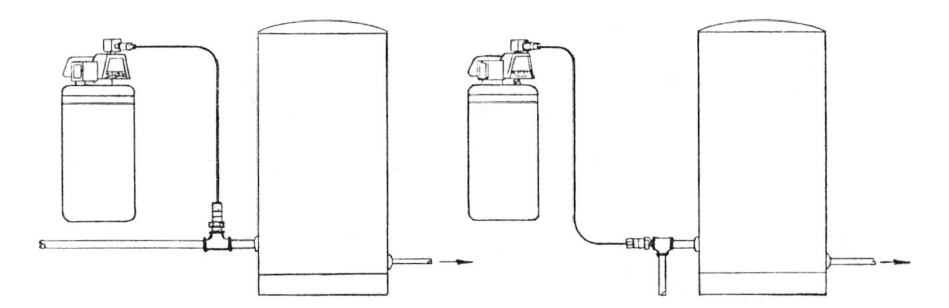

Figure 8-5. Proper injection installation. It is recommended that solution be injected *before* the water reaches the pressure tank to ensure complete mixing with water. (*Everpure, Inc.*)

hydraulic principle, employing the vacuum created when water flows either through a venturi tube or perpendicular to a nozzle. The vacuum draws the chlorine solution from a container into the chlorinator unit where it is mixed with water passing through the unit. The solution is then injected into the water system. In most cases, the water-inlet line to the chlorinator is connected so that it receives water from the discharge side of the water pump, and the chlorine solution is injected into the suction side of the same pump (Fig. 8-6). The chlorinator operates only

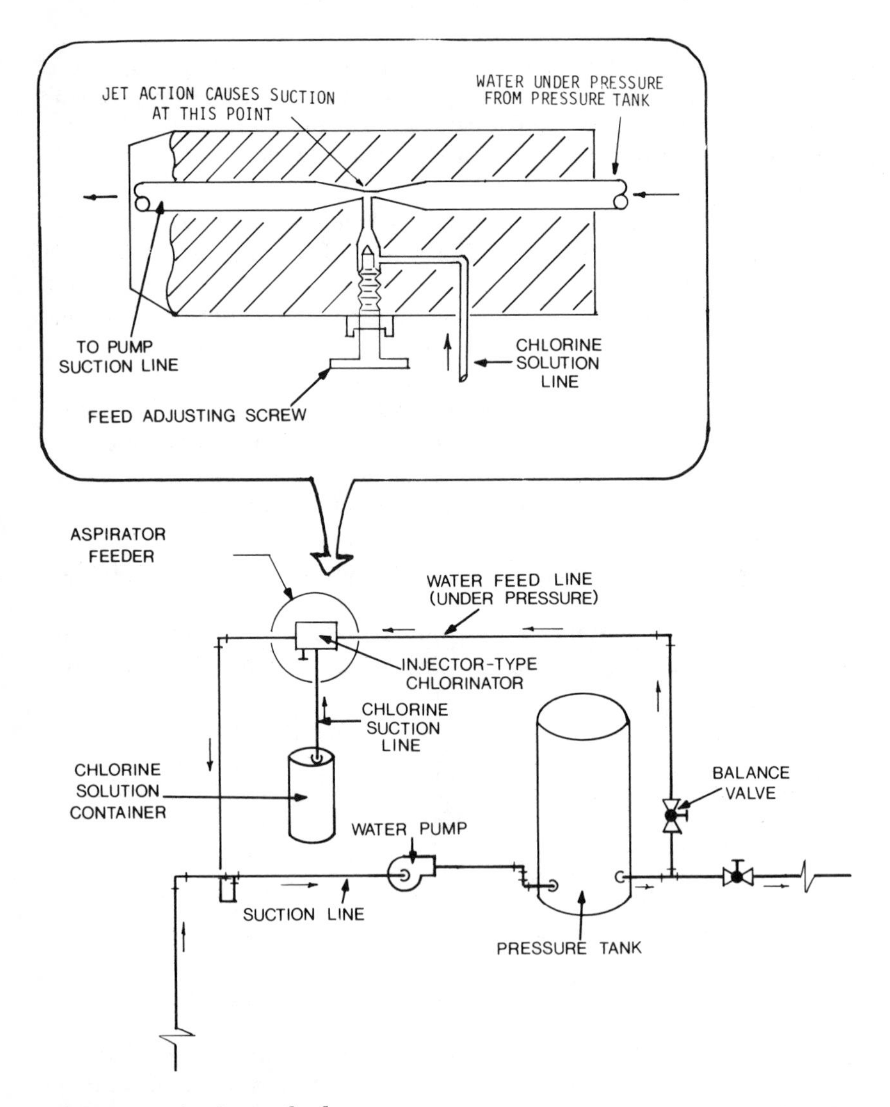

Figure 8-6. Aspirator feeder.

when the pump is operating. Solution flow rate is regulated by a control valve, although pressure variations may cause changes in the feed rate.

One type of suction feeder consists of a single line that runs from the chlorine-solution container through the chlorinator unit and connects to the suction side of the pump. The solution is pulled from the container by suction created by the operating water pump.

Another type of suction feeder operates on the siphon principle, with the chlorine solution being introduced directly into the well. This type also consists of a single line, but the line terminates in the well below the water surface instead of in the influent side of the water pump. When the pump is operating, its pump circuit should be connected to a liquid-level control so that the water-supply pump operation is interrupted when the chlorine solution is exhausted.

TABLET HYPOCHLORINATORS. The tablet hypochlorinating unit consists of a special pot feeder containing calcium hypochlorite tablets. It is accurately controlled by means of a flowmeter, by means of which small jets of feed water are injected into the lower portion of the tablet bed. The slow dissolution of the tablets provides a continuous source of fresh hypochlorite solution (Fig. 8-7a). This type of chlorinator is used when electricity is not available, and can operate where the water pressure is low.

It operates as follows:

1. The flow of water into the hypochlorinator is split into two streams, with the smaller of the two streams being diverted into the dissolving chamber.
2. During the operating phase, the depth of water in the dissolving chamber is kept at a constant value by a fixed-level overflow weir.
3. During the shutdown or nonoperating phase, a weep hole drains the water in the dissolving chamber to a level lower than the lowest possible position of the hopper, thereby stopping the dissolving or eroding of the disinfectant.
4. The rate of application of the disinfectant may be varied by adjusting the depth of submergence of the hopper in the dissolving chamber.

Although aspirator-type hypochlorinators have been marketed for some time, they have some inherent operation and maintenance problems, especially with feed rates. It is therefore recommended that

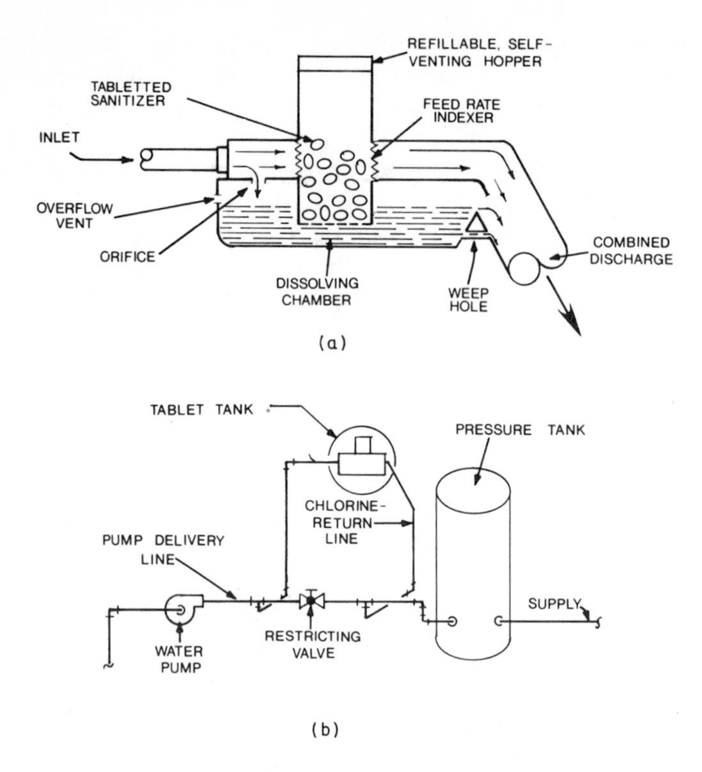

Figure 8-7. Tablet hypochlorinator.

positive-displacement chemical feeders (Fig. 8-7b) be used where conditions permit.

Determination of Proper Chlorine Dosage. Proper hypochlorination involves four basic factors: dosage, demands, residual, and contact time. *Dosage* is the amount of chlorine fed into the water system, expressed as milligrams of chemical per liter (mg/L). A set amount of chlorine, fed into the water, will oxidize or combine with chemicals such as ferrous iron, manganese, hydrogen sulfide, or nitrite, withdrawing them from availability during disinfection action. The amount of chlorine required is known as the *chlorine demand.*

The chlorine remaining in the water after the demand is filled, is known as the *residual.* If ammonia nitrogen is present in the water, some chlorine will combine with it to form chloramines, which have only mild germicidal capability. When there is no ammonia in the water, the remaining chlorine is called the *free-chlorine residual.* This has a 25-to-100-times greater disinfecting ability than do the chloramines.

Contact time is the period that elapses between the addition of

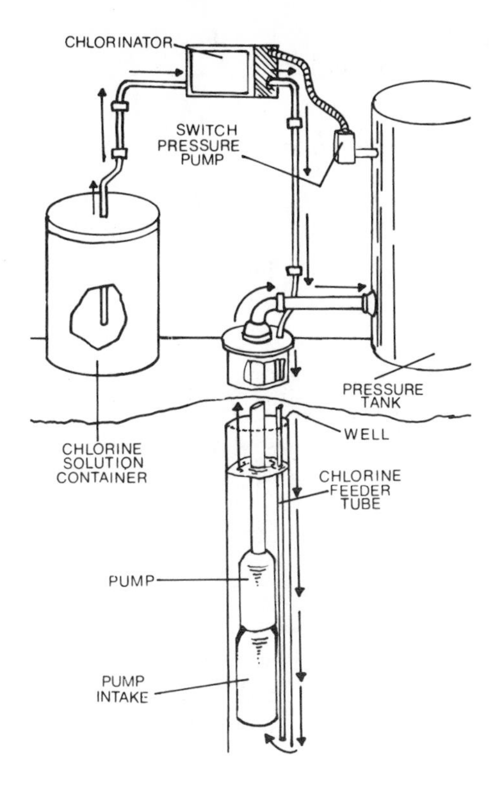

Figure 8-8.

chlorine and the water's use. Suitable contact time is required for the disinfecting action to occur. Contact times of 20 to 30 min are recommended, and the chlorine dosage should be great enough to provide a free-chlorine residual of 0.2 to 0.5 mg/L.

If a pressure tank does not provide adequate contact time for chlorination, the time of detention can be increased by adding a gravity tank or reservoir, by placing multiple tanks in series, or by adding lengths of tubing. In some cases contact time can be increased by feeding the chlorine into the well below the pump (Fig. 8-8). To avoid corrosion, the feed tube in a well should extend at least a foot below the pump assembly.

The primary factors that determine the biocidal efficiency of chlorine include the following:

1. Chlorine concentration: The higher the concentration, the more effective the disinfection and the faster the disinfection rate.

2. Type of chlorine residual: Free chlorine is a much more effective disinfectant than combined chlorine.
3. Contact time between the organism and chlorine: The longer the time, the more effective the disinfection.
4. Temperature of the water in which contact is made: The higher the temperature, the more effective the disinfection.
5. The pH of the water in which contact is made: The lower the pH, the more effective the disinfection.

Determination of Chlorine Residual. The practice of free-residual chlorination became widespread around 1939. This practice consists of adding enough chlorine to produce a residual made up almost entirely of free available chlorine. Because of its many advantages, including ease of control, the practice of free-residual chlorination is recommended for individual water-supply systems. If ammonia is present in the water, a free-chlorine residual can be obtained by adding sufficient chlorine to combine with all the ammonia nitrogen and form a compound known as nitrogen trichloride. Once this is done, the addition of any further chlorine will produce a free-chlorine residual.

Levels of free-chlorine residual can be easily and accurately determined by the DPD Colorimetric Method. The reagents required for a free-residual chlorine determination are contained in one tablet. When this tablet is added to a sample containing chlorine residual, a red color is produced that can be matched to a standard color comparator. Test kits for measuring free-residual chlorine levels of 0 to 10 mg/L are commercially available.

Advantages and Disadvantages of Chlorination. Chlorination is the most popular form of water disinfection practiced in America, and there is good reason for this. The chlorine residual that remains after water has left the disinfection unit provides continuing antibacterial protection. The amount of chlorine residual in the water can be readily determined with simple and inexpensive test kits. Chlorine is a highly effective disinfecting agent that is readily available at a reasonable price.

There are, however, some drawbacks to chlorination. Chlorinated organics are produced when certain organic materials combine with chlorine in water. Some of these chlorinated compounds are suspected of being carcinogenic. Because surface water contains higher concentrations of organic material than does ground water, the problem is suspected to occur more frequently in surface water than in ground water. Turbidity in the water can also reduce the effectiveness of chlorination. In addition, variations in water quality may affect the degree of bacteriological protection of the treated water. Because of these variations in effectiveness, doses often exceed necessary levels, producing high residuals which may impart a noticeable taste and odor to the water.

Finally, chlorination equipment should be properly maintained, as described in the customer's handbook usually provided by the water-conditioning dealer. Solution levels and residual-chlorine concentrations should be checked at least once a week.

Superchlorination–Dechlorination

In contrast to normal chlorination, superchlorination is the application of enough chlorine to maintain a minimum free-chlorine residual of 3.0 to 5.0 ppm, which is about 10 times higher than residuals resulting from standard chlorination practices. Although the idea of superchlorination is not new, it has not been commonly practiced. On a municipal water-treatment scale it is impractical because of the increased cost of the chemicals and activated carbon needed for dechlorination. This cost limitation is not applicable to domestic systems, however, because of the small volume of water treated in an individual system. The additional chlorine required for superchlorination during a year of operation of a household water supply would amount to only a few gallons of bleach. Dechlorination is needed only on water used for drinking and culinary purposes.

The advantages of superchlorination followed by dechlorination of individual water-supply systems are that

1. It provides more disinfecting activity than does normal chlorination.
2. The treatment technique compensates for the lack of control over widely varying factors which influence the effectiveness of chlorine disinfection.
3. It compensates for any lack of experienced supervision and timely maintenance of water-treatment and purification equipment.
4. The required contact time for disinfection is less than that for normal chlorination.
5. It supplies more palatable water for drinking and cooking by removing both undesirable tastes and odors and the excess chlorine during carbon filtration.

The equipment used in superchlorination and dechlorination is identical to that used for normal chlorination, except it supplies a higher dosage. The chlorinator treats the water before it reaches the pressure tank. The dechlorinator, usually an activated-carbon filter, receives the water after it has passed through the pressure tank (Fig. 8-9).

Ultraviolet Irradiation

Ultraviolet radiation's germicidal properties were recognized for many years before the first attempt was made to use it for disinfection of water in 1919. The killing action of ultraviolet light is the same as that provided by sunlight which kills bacteria in surface waters.

The ultraviolet water-disinfection system consists of one or more ultraviolet lamps enclosed in a quartz sleeve. Ultraviolet lamps produce electromagnetic radiation with a wavelength of 2537 angstroms (Å). (One centimeter, about 0.4 in, is equal to 100 000 000 angstroms.) The lamps are similar to fluorescent lamps, but they contain no phosphorescent coating on the inside of the tube to convert the ultraviolet radiation to visible light. The quartz sleeve surrounding each lamp protects it from the cooling action of the water, since the lamps must maintain a certain level of heat to produce the necessary killing effect.

The germicidal effect of ultraviolet radiation results from the intensity of the lamp and the length of exposure. The intensity of the light produced by a conventional germicidal lamp is sufficient to kill microorganisms in a fraction of a second. It is, of course, reduced with increasing distance from the lamp and by the medium through which it passes. In addition to the effect of turbidity, minute traces of iron compounds which commonly exist in water reduce the light transmission. For these reasons, ultraviolet water-purification units are designed so that water passes in a relatively thin layer around the lamps.

Each unit is designed (Fig. 8-10) so that the water flows at a particular rate. The flow must be regulated to meet this rate to ensure that all organisms present receive adequate exposure. The water should be filtered before it passes into the ultraviolet lamp chamber to eliminate the

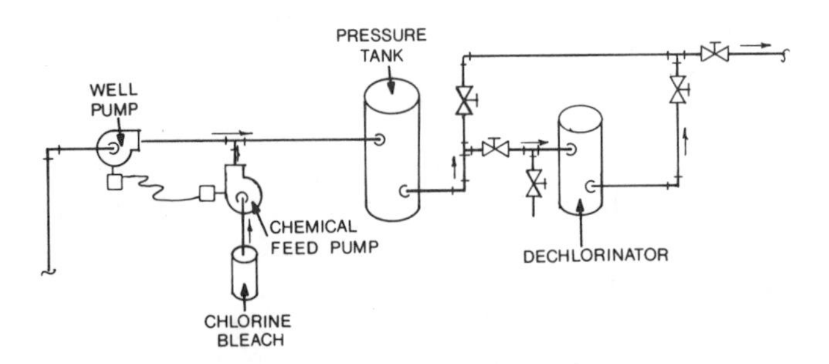

Figure 8-9. Chlorinator-dechlorinator.

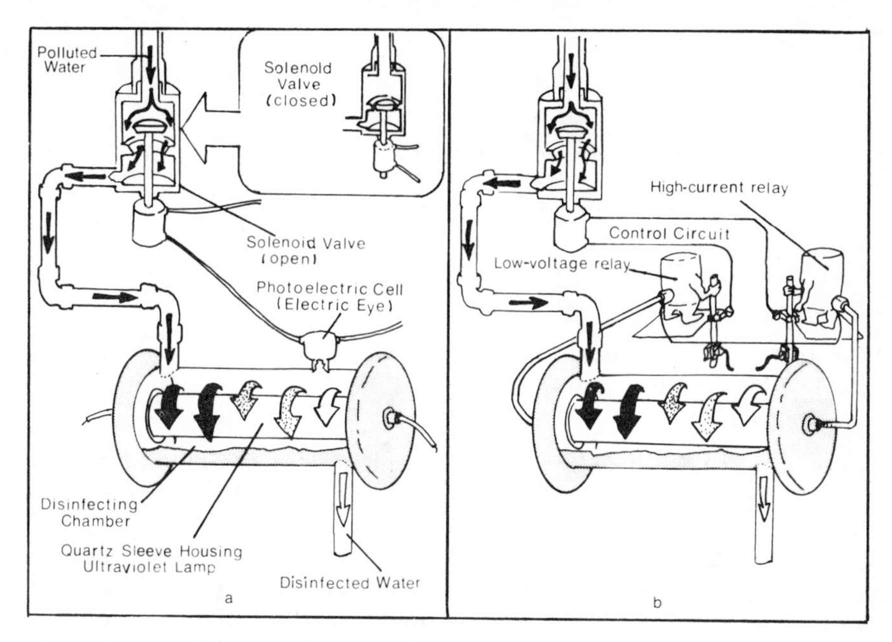

Figure 8-10. Ultraviolet water-purification unit. Types of automatic safety controls used for ultraviolet disinfection units. (*a*) Photoelectric cell holds the solenoid water valve open as long as the water passing through the disinfecting chamber is exposed to sufficient ultraviolet light. If dirt collects on the tube, if the tube breaks, if age limits the amount of light output, or if there is a power outage, the photoelectric cell closes the solenoid valve (inset). (*b*) Another type of ultraviolet control system provides protection with low-current and high-current relays. In case of no current or low voltage, the low-voltage relay opens the control circuit, and this causes the solenoid valve to close and shut off the water supply. As the lamp grows older and decreases in light intensity, the current (amperage) increases and reaches a point where the high-current relay opens the control circuit and shuts off the water supply by means of the solenoid valve.

possibility of an organism passing through in the shadow of a particle. The units are more effective when the flow is turbulent, because the organisms receive a more uniform exposure.

The intensity of the ultraviolet light emitted by the lamp gradually decreases as the lamp is used. Eventually the unit becomes relatively ineffective for killing bacteria. To offset this problem an ultraviolet-sensitive photocell is built in as a safety feature. This photocell will activate a solenoid valve to shut off the water if the intensity of radiation reaching the outside chamber is reduced below a minimum allowable level. The decrease in lamp output may be due to low or burntout lights, to sediment coating the wall of the tube, to highly turbid water,

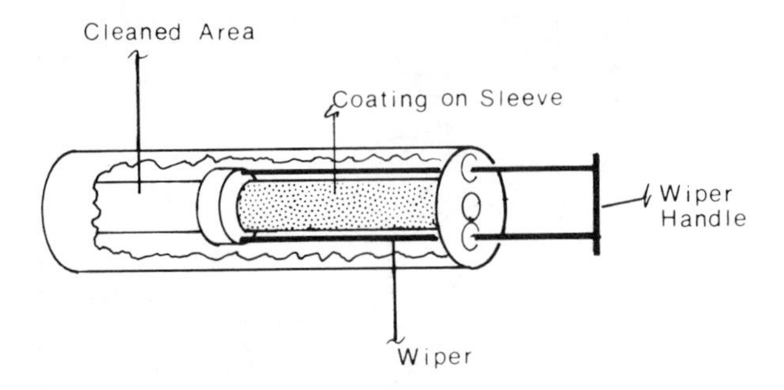

Figure 8-11. Hand-operated wiper for removing coating that forms on the quartz sleeve.

or to a power failure. For this reason, quartz sleeves around the lamp should be kept clean. Special devices are manufactured to clean the sleeves without removing them (Fig. 8-11).

Another safety device built into some ultraviolet-disinfection systems is a time delay that allows the lamp to reach its greatest output of ultraviolet light *before* the water system goes on.

Ultraviolet-disinfection units can be installed before or after the pressure tank, depending on the desirability of disinfecting all or just part of the water used in a home. A filter should be placed on the waterline before the ultraviolet unit, however, to remove particles of suspended matter. Ultraviolet light disinfection has several advantages:

1. The units are automatic, requiring a minimum of maintenance.
2. No undesirable materials are added to the water.
3. Minimal contact time is required.
4. No by-products are present to produce tastes or odors.

However, there are disadvantages:

1. The lamps have low penetrating power.
2. The effectiveness of the units can be reduced by turbidity or bacterial slime.
3. There is no simple test to determine the system's effectiveness.
4. The operational procedures leave no residual germicidal power in the water system.
5. The ultraviolet lamps gradually lose effectiveness.

Ozonation

Ozone is a form of oxygen having three atoms per molecule, rather than the two atoms typical of atmospheric oxygen. Ozone has greater germicidal effectiveness against bacteria and viruses than chlorine. It also reduces the iron, manganese, lead, and sulfur concentrations in water and eliminates most tastes and odors. Furthermore, its potency is not affected by pH, temperature, or ammonia content.

Ozone was first applied to water treatment in the late 1800s. Although over 1000 cities worldwide use ozone for water disinfection, it is not common in America. This reflects an American desire to maintain residual germicidal power in the distribution system, and to avoid the higher equipment and operating costs of an ozonation system. However, the future of ozonation for water purification in individual systems may be bright. Since there is no extensive distribution system, there is little need for maintaining a germicidal residual.

Because ozone (O_3) is an unstable molecule, it quickly reverts to normal oxygen (O_2). Therefore, ozonation must occur at the point of use. The most practical way of generating ozone is to pass oxygen through a corona discharge (Fig. 8-12), produced by applying high voltage across two electrodes with a dielectric (insulating sheet) and an air gap between. The dielectric converts the normal electric arc into a blue glow of electrons. The electrons break the O_2 molecules into atoms, some of which recombine as O_3, or ozone. The electrodes and dielectrics are usually arranged as parallel plates or concentric tubes. Much of the electric power used turns to heat, so the electrodes must be cooled by air or liquid. The ozone content of the gas produced by the ozone generator depends on the oxygen and moisture content, as well as on the flow rate of the feed gas and the applied voltage. Because air's

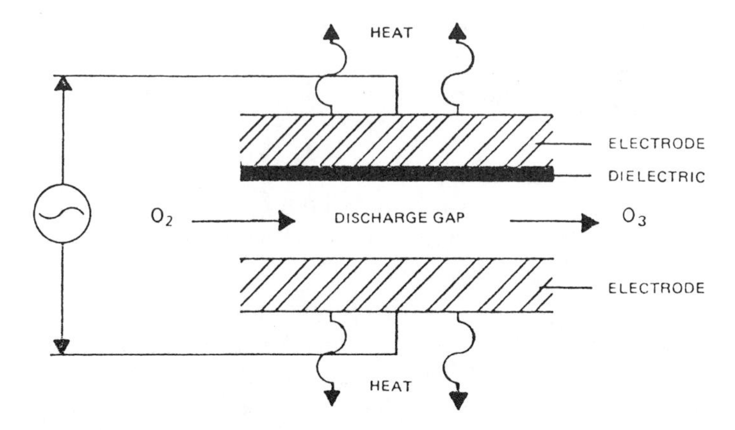

Figure 8-12. Basic ozonator configuration.

moisture content reduces the performance of the ozone generator, it may be necessary to dry the air before it passes through.

The water to be purified passes through a venturi throat creating a vacuum which then sucks the ozone gas into the water. Where water hardness is a problem, polyphosphate additives can prevent carbonate precipitation. Some have also claimed that polyphosphates improve the distribution of ozone in the water.

There are a number of advantages to using ozone as a disinfectant, as follows:

1. Because ozone purifies naturally with a form of oxygen, no chemical residual persists.
2. Ozone is the strongest germicide available for use. It rapidly deactivates microorganisms (including viruses).
3. Ozonation also eliminates problems of taste and odor in the water supply.
4. As long as there is an excess of ozone, the dosage is self-regulating, and excess ozone will rapidly dissolve from the water. Therefore, variations in ozone demand can normally be met automatically.
5. Ozone is effective over a wide pH range, so pH adjustment is seldom necessary.

The disadvantages of ozonation include the following:

1. Higher equipment and operating costs than chlorination.
2. Unreliability of equipment. In an EPA study in Grand Isle, Iowa, the ozonation equipment required frequent and varied maintenance and repair after less than a year of operation.
3. There is no residual germicidal power left in the distribution system. It should, however, be pointed out that the effectiveness of the ozonation can be measured immediately after the point of treatment.
4. The effects that many of the by-products of ozonation have on human health are still unknown. It is possible that some types of new carcinogenic substances may be produced. More research will have to be done before more definitive conclusions can be drawn.

Silver

It has been known for some time that silver is a good bactericide. The germicidal action of silver, like that of metals such as mercury, copper, or zinc, is thought to result from its reaction with the proteins in cell protoplasm.

Although there is some disagreement in the literature, the effectiveness of silver appears to occur regardless of the method by which it is introduced into the water. In commercially available silver-treatment units, the water usually dissolves the silver from a diatomaceous-earth or activated-carbon filtering medium which has been impregnated with silver nitrate or some other form of silver. The activated-carbon filter also removes tastes and odors from the water.

Silver-treatment units range in size from a couple of inches in diameter and a foot long to 6 or 7 inches in diameter and 4 or more feet long. The larger units do not necessarily have a higher capacity, but they can be backwashed when necessary to restore the unit to its designed flow rate. Because each unit is designed for a maximum flow rate, a flow regulator should be used to ensure that the design flow rate is not exceeded.

Concentrations of silver as low as 15 parts per billion (ppb) are sufficient to destroy most microorganisms, if given enough time (Fair, 1971). U.S. EPA's drinking-water standards require that the maximum concentration of silver in water not exceed 0.05 ppm. Most silver-treatment devices raise silver concentrations in water to only 0.03 to 0.04 ppm. At these concentrations, silver imparts no taste to water.

Although silver is expensive, a pound of silver could provide a dosage of 0.035 ppm to about 3½ million gallons of water. Therefore, the cost of treating water with silver is relatively inexpensive.

Advantages of silver-disinfection systems include the following:

1. Simplicity. The process is automatic and the system has no moving parts to adjust. The only maintenance required is replacement and/or backwashing of the filter.
2. Disinfection with silver imparts no taste or odor to the water.
3. The disinfection units are small and can easily be installed in the waterline.

The limitations of using silver for waste disinfection are as follows:

1. The filters have a fixed capacity.
2. If the water is turbid and contains suspended sediment, the filters will plug rapidly and will require prefiltered water, frequent cleaning, or replacement.

3. There is no simple residual test by which the homeowners can determine whether they are getting the correct amount of silver in their water.
4. Silver has a tendency to react with organic matter, iron, sulfur and other chemicals, which inhibits its bactericidal action.
5. The contact time required for silver is much greater than that for chlorine, ranging from minutes to hours for complete disinfection.

Because of the problems of interfering chemicals and the long contact time required for complete disinfection, use of silver as a germicide for an individual water-supply system is not normally advisable. In fact, the federal government authorizes the use of silver only for the treatment of water *already* defined as potable.

Iodination

Iodine is a chemical oxidant which disinfects in a manner similar to that of chlorine. It is not likely that iodine will ever replace chlorination as the most common method of disinfection because of its higher cost. However, there are many situations where iodination may be more practical than chlorination or other methods of disinfection.

Iodine is the least soluble of all the halogens and, therefore, the least likely to be hydrolyzed by water. Iodine also has the lowest oxidation potential, reacting more slowly with organic compounds than chlorine. Since iodine is chemically more stable, it does not react with nitrogenous compounds as does chlorine.

Because of the stability of iodine and the simplicity of its hardware, iodine disinfection units were used by the National Aeronautics and Space Administration in all lunar modules to protect the drinking water of the astronauts during flights. Iodination has also played an important role in emergency disinfection of the drinking water for the U.S. Armed Forces. The U.S. Forest Service uses iodination equipment on many of its water wells which supply potable drinking water to remote areas because the equipment has no moving parts and requires no electricity to operate.

As a disinfectant, iodine remains effective through a wider pH range than does chlorine: while the effectiveness of chlorine starts to drop off after a pH of 8, iodine does not start to lose its bactericidal effect until the pH of water reaches 10.

The effective bactericidal forms of iodine are elemental iodine (I_2)

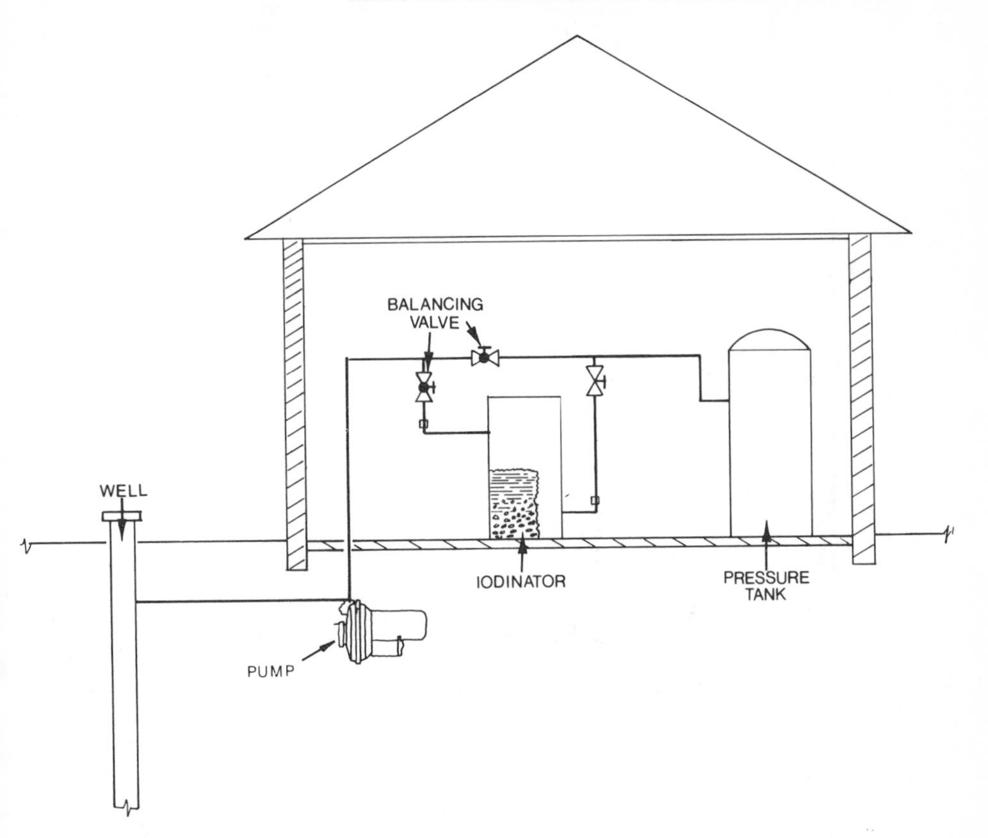

Figure 8-13. Iodinator.

and hypoiodous acid (HOI). Other forms of iodine are not considered to have a killing effect on pathogenic bacteria.

Iodination equipment is installed between the pump and a holding tank (or pressure tank) and as near to the pump as possible (Fig. 8-13). To provide an adequate contact time for the iodine's germicidal activity to take place, some type of detention tank will be needed if there is no pressure tank (Table 8-2). The iodination equipment feeds a continuous flow of concentrated iodine into the main stream of the water. The iodine solution is injected by a diverted flow of water caused by back pressure in front of the iodination system. This diverted water flows through the iodination tank, which is filled with iodine crystals.

Because iodine concentration is directly related to ambient temperatures, a saturated iodine solution of approximately 300 ppm at 20°C is then injected into the main water flow. Iodine residuals of 0.5 mg/L to 1.0 mg/L should persist for at least a half hour after injection. Adjust-

Table 8-2. Approximate Contact Time in Which These Potentially Dangerous Organisms Can Be Controlled with an Iodinamics Iodinator® *(0.5 ppm iodine, pH 7.5, 20–26°C)*

Organisms	*Disease caused*	*Approximate contact time*
Bacteria:		
Escherichia coli	Cystitis of urinary tract	50 s*
Salmonella typhosa P-4	Typhoid fever; gastroenteritis	1 min†
Salmonella typhosa P-5	Typhoid fever; gastroenteritis	1 min†
Salmonella typhosa P-10	Typhoid fever; gastroenteritis	1 min*†
Salmonella paratyphi P-2	Paratyphoid fever	1 min†
Salmonella schottmuelleri P-3	Paratyphoid fever	2 min†
Salmonella typhimurium P-6	Food poisoning	5 min†
Shigella flexneri P-7	Paradysentery	2 min†
Shigella dysenteriae 11 P-8	Dysentery; intestinal ulcers	2 min†
Shigella sonnei P-9	Paradysentery	2 min†
Streptococcus fecalis E-40	Can be pathogenic	2 min†
Staphylococcus aureus	Septicemia; brain abscess; enteritis	50 s‡
Staphylococcus epidermidis	Subacute endocarditis	1 min‡
Virus:		
Poliovirus type 1	Polio	9 min*
Cysts (@ 1 ppm iodine):		
Entamoeba histolytica	Severe dysentery	30 min*

Contact times to control the above bacteria, viruses and cysts have been obtained through the research of the following persons and their reports upon which we are relying. The listing of these times can be found in the cited publications.

* Black et al., "Iodine for the Disinfection of Water," *J. A.W.W.A.*, vol. 60, no. 1, January 1968.

† C. W. Chambers et al., Bacteriology Section, Environmental Health Center, U.S. Public Health Service, Cincinnati, Ohio.

‡ M. A. Keirn, and H. D. Putnam, *Health Laboratory Science*, vol. 5, July 1968.

ments in iodine concentrations can be made by altering the amount of back pressure and the flow of water through the iodination unit. The residual can be determined using a simple test kit.

The overall system should be designed to permit an approximate contact time of 15 min, so that pathogenic organisms can be effectively controlled. The length of the contact time is dependent on the relative activity of the cell wall of the organism. Thus, actively metabolizing microbes require less contact time than do cystated microbes.

The advantages of iodination disinfection are the following:

1. Its bactericidal capacity is not greatly influenced by pH except at very low temperatures.
2. Ammonia and organic nitrogenous impurities have little effect on disinfection efficiency because they do not form substitution compounds with iodine.
3. Its action depends less on contact time and temperature than does chlorine's.
4. It is highly effective against pathogenic organisms (including spores, viruses, etc.) in short contact times.
5. Iodination equipment requires little maintenance and its operation is very simple.

Despite its effectiveness and advantages, iodination has never become an important water-disinfection technique. Its lack of popularity may be due to the limitations listed below.

1. Iodination requires higher dosages for comparable disinfection activity under most conditions, except when the pH of water is higher than 8.
2. Iodine is expensive—about 20 times more expensive than chlorine per unit of germicidal effectiveness.
3. Iodine may impart a taste or a slight color to water thereby affecting its palatability and aesthetic appeal.
4. The large scale preparation of stock iodine solutions is currently impractical and economically prohibitive.
5. Tests have been run with sample populations that disinfected their water by iodination for periods of up to a year. Although participants in such tests experienced no physiological symptoms, the effects of prolonged use, especially on children, remains to be ascertained. Evidence also indicates

that some people will experience an allergic reaction resulting from iodine ingestion.

6. If iodination is to be used to destroy cysts and viruses, more attention must be given to maintaining a near-neutral pH level.

Pasteurization

Water pasteurizers are a relatively new approach to disinfecting individual water supplies. Pasteurization is a process by which the water temperature is raised above 140°F and maintained until all pathogenic bacteria are destroyed. The higher the temperature, the shorter the required detention time.

Figure 8-14 is a schematic diagram of one type of water pasteurizer that was developed at the Robert A. Taft Sanitary Engineering Center, Cincinnati, Ohio (Goldstein, 1960). Untreated water enters the heat exchanger, where its temperature is raised to about 151°F by exchange of heat from the treated water flowing countercurrent to it. A bypass pump sends the water from the heat exchanger through an electric heater, which raises the temperature to this system's pasteurization point of 161°F. A 30-ft ¼-in (914.40-cm 6.35-mm) copper detention tube assures holding time of at least 15 s at that heat. An air-relief valve permits the gases liberated at the higher temperature to escape.

If the water temperature falls below 161°F after the water passes through the retention tube, a solenoid valve and thermostatic switch are designed to return the water to the suction side of the bypass pump and to recycle it through the heater. When the water at the thermostatic switch is 161°F or higher, it flows to the heat exchanger, where it gives up its warmth to the incoming water. The water is thus cooled to within about 10°F of the temperature of the incoming water before it enters the storage tank. A pressure pump also provides water to the household plumbing system unless the storage tank is at an elevation sufficient to supply water by gravity flow.

The auxiliary line shown in Fig. 8-14 supplies water for priming the system. The line, which consists of a hose or other temporary connection, should not be connected permanently to the system.

Pasteurization represents a satisfactory method of destroying the pathogens that drinking water can contain. In emergency situations when contamination has possibly occurred, heat has been successfully used for disinfection. Overall operation of the system is, moreover, both reliable and simple.

Pasteurization of water is very expensive, however. The total cost of water pasteurization on a household scale is estimated to exceed $1 per

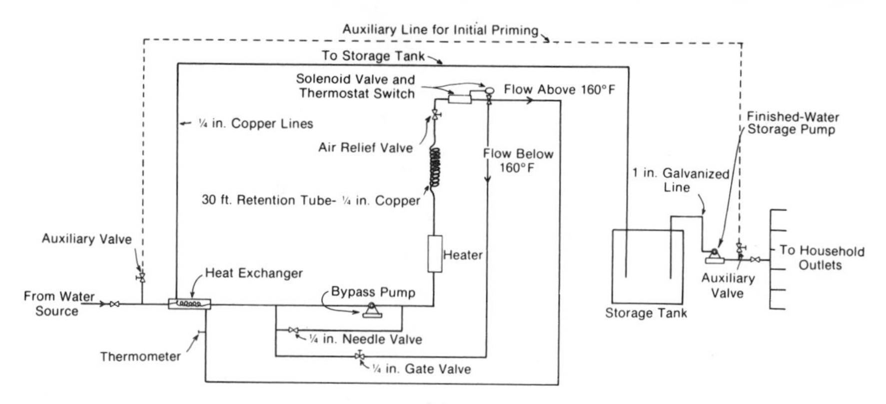

Figure 8-14. Flow diagram of a continuous-flow water pasteurizer.

1000 gal (3785 L). Much of this cost pays for the energy required by the system. To perform even somewhat economically, the pasteurization unit must operate during at least 12 h each day. When the unit operates less, the heating-up and cooling-off periods greatly reduce its efficiency.

Other disadvantages of pasteurization include a limitation on the quantity of water that can be disinfected at any one time, as well as the unit's inability to provide residual germicidal protection.

DISCUSSION

Chlorination will probably continue to dominate over the described water-disinfection alternatives. If homeowners decide to select a disinfection system other than chlorination, they should first check with state and local health officials to ensure that such treatment of water does not conflict with any local or regional drinking-water regulations.

REFERENCES

Baumann, E. R., and D. D. Ludwig, 1962: "Free Available Chlorine Residuals for Small Nonpublic Water Supplies," *J. A.W.W.A.*, vol. 58, no. 11, November, pp. 1379–1388.

Breek, R. S., E. G. D. Murray, and N. R. Smith, 1957: *Bergey's Manual of Determinitive Bacteriology*, 7th ed., Williams and Wilkins, Baltimore.

Center for Disease Control, 1976: "Foodborne and Waterborne Disease Outbreaks," Annual Survey, 1975. U.S. Department of Health, Education, and Welfare Publication (CDC) 76-8185.

Craun, G. F., L. J. McCabe, and J. M. Hughes, 1976: "Waterborne Disease Outbreaks in the U.S. 1971–1974," *J. A.W.W.A.*, vol. 68, no. 8, August, pp. 420–424.

Everpure, Inc., 1962: *Water Purification and Treatment Handbook*, Oak Brook, Illinois.

Fair, G. M., et al., 1971: *Elements of Water Supply and Waste Water Disposal*, Wiley, New York.

Geldreich, E. E., 1972: "Water Borne Pathogens," in *Water Pollution Microbiology*, Ralph Mitchell (ed.), Wiley-Interscience, New York.

Goldstein, M., et. al., 1960: "Continuous-Flow Water Pasteurizer for Small Supplies," *J. A.W.W.A.*, vol. 52, no. 2, February, pp. 247–254.

Hoehn, R. C., 1976: "Comparative Disinfection Methods," *J. A.W.W.A.*, vol. 68, no. 6, June, pp. 302–308.

Hoye, R., G. S. Logsdon, and J. M. Symons, 1978: "Removal of *Giardia* Cysts and Cyst Models by Diatomaceous Earth Filters," unpublished report, U.S. EPA, Cincinnati, Ohio.

Johnson, J. Donald (ed.), 1975: *Disinfection—Water and Wastewater*, Ann Arbor Science, Ann Arbor, Michigan.

Johnson, J. D., and R. Overby, 1971: "Bromine and Bromamine Disinfection Chemistry," *J. Sanit. Eng. Div., Proc. Amer. Soc. Civ. Eng.* vol. 97 (SA5), October, pp. 617–628.

Kruse, C. W., 1969: "Mode of Action of Halogens on Bacteria, Viruses and Protozoa in Water Systems," Final Report to U.S. Army Medical Research and Development Command, Washington, D.C.

Rice, R. G., and J. A. Cotruvo (eds.), 1978: *Ozone/Chlorine Dioxide Oxidation Products of Organic Materials*, Ozone Press International, Cleveland, Ohio.

Seiden, Rudolph, 1968: *Livestock Health Encyclopedia*, Springer, New York.

Sharp, D. G., and J. D. Johnson, 1977: "Inactivation of Viruses in Water by Bromine and its Compounds: Influence of Virion Aggregation," Final Technical Report, U.S. Army Medical Research and Development Command Contract DAMD-17-74-C-4013, February.

Symons, J. M. (ed.), 1977: *Ozone, Chlorine Dioxide, and Chloramines as Alternatives to Chlorine for Disinfection of Drinking Water*, U.S. EPA, Cincinnati, Ohio, November.

U.S. Environmental Protection Agency, 1973: *Manual of Individual Water Supply Systems*, U.S. EPA Office of Water Programs, Water Supply Division, EPA-430/9-74-007, Washington, D.C.

9

Treatment Techniques for the Removal of Chemical Contaminants

HARDNESS

Hardness enters a water supply when calcium and magnesium salts are dissolved by ground water. It can be present in many forms, the bicarbonate, sulfate, and chloride salts being the most common. Each will cause its own form of trouble.

Carbonate hardness is the result of rain water dissolving limestone, i.e., calcium and magnesium carbonate. This begins when water dissolves carbon dioxide gas to form carbonic acid:

$$\underset{\text{Water}}{H_2O} + \underset{\text{Carbon dioxide}}{CO_2} \rightleftharpoons \underset{\text{Carbonic acid}}{H_2CO_3}$$

This weakly acidic water is aggressive and will tend to dissolve many minerals with which it comes in contact. When it dissolves limestone, it forms solutions of calcium and/or magnesium bicarbonate:

$$\underset{\substack{\text{Carbonic acid in}\\ \text{Rain water}}}{H_2CO_3} + \underset{\substack{\text{Calcium carbonate in}\\ \text{Limestone}}}{CaCO_3} \rightarrow \underset{\substack{\text{Calcium bicarbonate in}\\ \text{Hard water}}}{Ca\,(HCO_3)_2}$$

$$H_2CO_3 + \underset{\text{Magnesium carbonate}}{McCO_3} \rightarrow \underset{\text{Magnesium bicarbonate}}{Mg\,(HCO_3)_2}$$

This form of hardness is unstable, and when exposed to heat, will revert to its original limestone state.

$$\underset{\substack{\text{Calcium}\\ \text{bicarbonate}}}{Ca(HCO_3)_2} \xrightarrow{\text{Heat}} \underset{\substack{\text{Calcium}\\ \text{carbonate}}}{CaCO_3} + \underset{\text{Water}}{H_2O} + \underset{\substack{\text{Carbon}\\ \text{dioxide}}}{CO_2}$$

$$\underset{\substack{\text{Magnesium}\\ \text{bicarbonate}}}{Mg(HCO_3)_2} \xrightarrow{\text{Heat}} \underset{\substack{\text{Magnesium}\\ \text{carbonate}}}{MgCO_3} + \underset{\text{Water}}{H_2O} + \underset{\substack{\text{Carbon}\\ \text{dioxide}}}{CO_2}$$

The calcium and magnesium carbonates are relatively insoluble and readily deposit on heat exchangers (i.e., heating elements in a home water heater). Because of instability with heat and scaling tendencies, calcium and magnesium bicarbonates are often referred to as "temporary" hardness. The scale formed in a water heater insulates the heating elements and surfaces of a water heater, greatly reducing fuel efficiency. It has been stated that ⅛ in (3.18 mm) of scale can increase fuel consumption of a water heater by up to 30 percent. When deposited in sufficient amounts, this scale can cause localized overheating which can "burn out" the heater.

Of the sulfate hardness compounds, magnesium sulfate has a higher solubility and is more common. The sulfate salts can add to the scale problem but are more often associated with the laxative effect they can have on human beings and livestock. Indeed, magnesium sulfate, commonly called Epsom salt, is one of the most common natural laxatives. Sulfate hardness has cost livestock producers dearly in death or stunted growth of cattle and swine. The sulfate compounds also cause water to taste bitter or strong. Calcium sulfate will cause scale to precipitate from waters that are heated very high.

The last of the common hardness salts are calcium and magnesium chloride. Chloride hardness occurs when highly soluble calcium and magnesium chloride are dissolved by ground water. Because of their high solubility, chloride salts do not cause a scale problem, but they are not without their bad traits. They cause a brackish or salty taste and can contribute heavily to the corrosive tendency of water. The chloride and sulfate hardness are referred to as permanent hardness because they tend not to precipitate as scale.

Hardness in any form will combine with soap to form an insoluble curd. Hardness problems are solved quite simply and economically. Ion-exchange water softeners are inexpensive and easily installed and maintained.

Simplified, an ion-exchange softener is a tank containing a bed of insoluble material. This material has a negative charge and has positive sodium ions "attached" to it. The resin has a stronger affinity for calcium and magnesium ions than it does for sodium. Thus, when hard water containing calcium and magnesium ions passes through the resin, the hardness ions are attracted to the resin, thereby releasing the sodium ions in an equivalent quantity to the water supply. In essence, the water softener trades sodium ions for hardness ions—hence the term ion exchange. There is no change in the total ionic content in the water. Before all the sodium ions have been displaced, hardness appears in the treated water. The resin is now said to be exhausted and must be regenerated. Regeneration is accomplished by passing a strong (½-saturated strength) sodium chloride solution through the resin. The high concentration of sodium ions overcomes the natural affinity for hardness so that the softening process is reversed. Sodium ions are placed on the resin while hardness ions are washed to the drain with the spent brine. The exhaust-regenerate cycle can be repeated almost indefinitely without damage to the resin.

Modern water softeners contain a plastic resin that has been specially activated. The resin is manufactured by copolymerizing styrene and divinyl benzene; the compounds cross-link to form a tough, durable polystyrene bead. The particle size of the beads ranges from 0.3 to 1.2 mm, with 97 percent falling in the 0.4- to 0.8-mm-size range. After polymerization, sulfonation converts the inert plastic resin to a chemically active ion exchanger by adding a sulfonic acid radical to the copolymer structure:

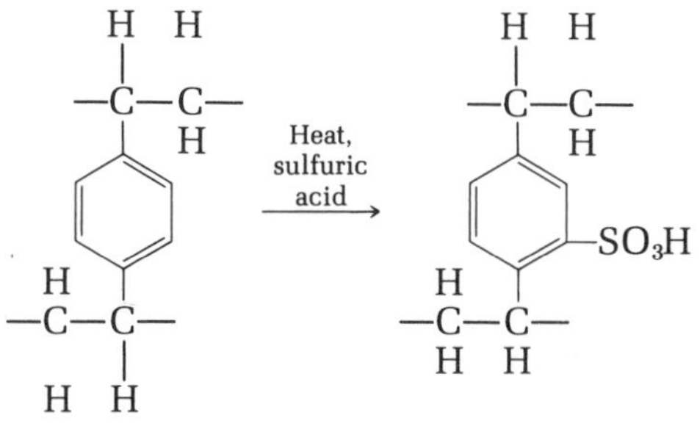

```
     H  H                              H  H
     |                                 |
   —C—C—                             —C—C—
     |  H        Heat,                 |  H
    ___          sulfuric             ___
   /   \          acid               /   \
  |     |      ——————————→          |     |—SO3H
   \___/                             \___/
H    |                          H      |
—C—C—                           —C—C—
     |                          H  H
H    H
```

To simplify matters, this resin will be referred to as $H \cdot SO_3R$. The hydrogen on the resin is the exchangeable ion and the sulfonate group is the active radical. The final step in preparing the resin for use in a softener consists of rinsing the resin with sodium carbonate:

$$2H \cdot SO_3R + Na_2CO_3 \rightarrow 2Na \cdot SO_3R + CO_2\uparrow + H_2O$$

The resin is now in the sodium form; after a good rinse with clean, soft water, it is ready to use.

The simple form of an ion-exchange water softener is a vessel containing ion-exchange materials. This system is used in the rental/exchange business, where the softener is removed from the installation for regeneration at a central facility.

Household automatic softener will usually consist of three basic components: the valve/control mechanism, a brine tank, and a resin tank. The resin tank is the workhorse, the valve/controller is the brains, and the brine tank, the maintenance apparatus. The mechanics of the softener will be discussed in Chap. 11. We will focus here on the work of the resin tank.

The average household water softener contains approximately ⅓ to 1½ ft^3 of resin. The total capacity of a softener, then, is largely depen-

Table 9-1. Capacity versus Salt Dosage

Resin volume, ft³	*Grains capacity/pound salt*		
0.3	10 000/5	8700/3	6700/2
0.5	15 000/8	13 000/5	10 000/3
0.7	21 000/10	17 300/7	13 300/4
1.0	30 000/15	26 000/10	20 000/6
1.5	45 000/23	39 000/15	30 000/9

dent on the volume of resin it contains. A second, and very important factor is the amount of salt used to regenerate the softener. In well-designed units, the values given in Table 9-1 would be typical.

Based on Table 9-1, one could conclude that 40 percent of the salt accomplishes 60 percent of the regeneration. Indeed, it is wise to regenerate a softener more often at a lower salt dosage.

The maximum efficiency of resin can be realized only in a properly designed softener. Consideration must be given to depth of resin bed, service and regeneration flow rate, brine concentration, and salt dosage per bed volume. Table 9-2 gives the recommended operating parameters. The typical cycle of a water softener includes softening, backwash, brining, and rinse.

Of the factors involved, salt dosage per cubic foot of resin is the most significant to the unit's efficiency. Figure 9-1 demonstrates this fact. As the salt dosage increases, the salt efficiency decreases dramatically. Generally it is recommended that the minimum dosage per cubic foot of resin be 5 lbs (2.25 kg). This is necessary to prevent hardness bleed during the service cycle.

The size of softener required by a particular application is dependent on three things: the hardness level of the supply, the amount of water to be used daily, and the flow rate of the system.

Table 9-2. Ion-Exchange Resin-operating Parameters

Bed depth	24–36 in
Service flow rate	2–10 gpm/ft³ of resin
Backwash flow rate	5–6 gpm/ft³
Brine concentration	10–16% NaCl
Regenerant contact time	35–60 min
Regeneration flow rate	0.2–1.0 gpm/ft³
Rinse flow rate	1–5 gpm/ft³
Rinse requirement	20–40 gal/ft³

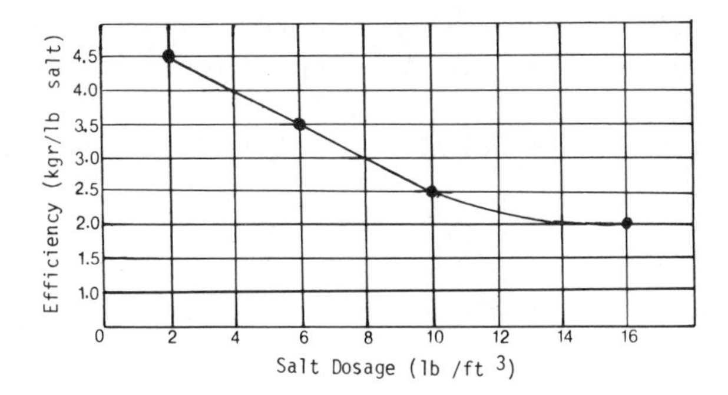

Figure 9-1. Salt dosage vs. efficiency (1 ft^3 ion-exchange resin).

Effect of Water Hardness, Usage, and Flow Rate

It would seem obvious that the higher the level of hardness in a supply, the larger the softener required. The installation of a small softener on extremely hard water would result in hardness bleed or leakage during the service cycle. Also, the larger the daily demand, the larger the softener: This is necessary to assure sufficient capacity between regenerations. Flow rate can affect the performance of the softener, and a softener of adequate size must be used to prevent hardness leakage during service.

Hardness is determined by testing the water. This can be done by a laboratory, either a local water-analysis lab or a manufacturer's lab. Often manufacturers provide this service at no charge to the consumer. Even though sales representatives can also test water hardness, it is usually a good idea to have the results verified by a laboratory.

The water usage can be estimated by the homeowner. If all water used in the home is to be softened, a factor of 60 gal/day (230 L/day) for each member of the household is allowed.

Example: 5 persons in household
$5 \times 60 = 300$ gal/day (1150 L/day)

If only the hot water is to be softened a factor of 25 gal/(person)(day) [950 L/(person)(day)] is used. This household would need 125 gal (475 L) of hot water each day.

The potential flow rate of the system depends on the type and number of water-using fixtures and appliances in the home. However, since the flow rate of the system cannot exceed the capacity of the pump, the pump's specified flow rate can safely be used in selection of the softener size.

Following is an example of the calculations and considerations which must be made when choosing a water softener for a typical household of four. The water will be used for general domestic purposes and no special treatment is required. The well system produces water at a rate of 10 gal/min (40 L/min) and its quality is relatively good: it is free of contaminants except for 20 gpg of hardness. Floor drains, electric outlets, shelter, and floor space are all adequate.

1. Daily usage requirements:

   4 people × 60 gal/(person)(day) = 240 gal/day (915 L/day)

2. Daily hardness removal:

   240 gal/day × 20 gpg = 4800 gr/day

Because the flow rate is 10 gal/min, a 1 = ft^3 softener should be used. Based on the daily requirement of 4800 gr, the regeneration frequency can be determined: 30 000 gr/15 lb salt × 4800 gr/d = 6 days.

This would require a regeneration once every 6 days at a 15-lb (6.8 kg) salt dosage. But what if we were to choose one of the more efficient dosages as outlined in Table 9-3.

The facts are obvious. By regeneration every fourth day at a lower salt setting, the annual salt consumption is reduced by 40 percent: The consumer could realize a savings of up to $20/yr. Some automatic softeners are available with an electronic sensor that will initiate the regeneration only when it is needed. This overcomes the inefficiencies associated with averaging the water used [60 gal/(person)(day)], varying water hardness, and variable water usage. An analogy has been made that such softeners are like the automatic water heaters that heat the reservoir water when it has been replaced with cold water.

Most manufacturers list the recommended hardness and flow-rate limits of their equipment. When this information is not available the parameters given in Table 9-4 are suggested. The variation in the upper hardness limit depends on the daily usage and the flow rate of the system.

Table 9-3. Annual Salt Usage at Various Dosages

Salt dosage per regeneration, lb	*Available capacity, gr*	*Regeneration frequency*	*Regenerations per year*	*Annual salt consumption, lb*
6	20,000	Every fourth day	91	550
10	26,000	Every fifth day	75	730
15	30,000	Every sixth day	61	915

Table 9-4. Hardness Limits

Bed volume, ft³	*Capacity, 15 lb/ft³; gr*	*Maximum hardness, gpg*	*Maximum recommended flow rate, gal/min*
0.3	10 000	20	3–4
0.5	15 000	25–30	5–6
0.7	21 000	40–50	7–8
1.0	30 000	60–70	10–12
1.5	45 000	75–90	15–18

As always, there are limitations that determine how efficiently the equipment will perform. These must be taken into account when planning a water-purification system. In some instances it may not be advisable to soften an entire water supply, for example, when it contains high levels of dissolved solids. These will impart a definite taste to water. But the hardness ions are exchanged for sodium ions when the water is softened, so the treated water tastes very different from the untreated water. Frequently users may find it difficult to acclimate to the different-tasting water. It is usually easy to provide an unsoftened tap to the kitchen for drinking water if the consumer desires. Another alternative would be to provide a small still or reverse-osmosis unit for further treatment of the drinking water. It should be noted that all dissolved minerals will cause a taste if present in high-enough quantities. Typically though, a consumer can become accustomed to the taste of dissolved minerals within a short time.

A water supply that contains a high concentration of dissolved minerals can directly inhibit the ion-exchange softening process and so may necessitate modifications in the treatment process. As pointed out above, an exhausted softener can be regenerated with a strong solution of sodium ions. If a water supply contains a substantial amount of sodium, along with the hardness, the sodium produces a weak regenerative effect, which will prevent the complete removal of the hardness. Some hardness will thus "leak" through the softener.

A similar effect is produced when the water hardness is very high, even if the sodium in the hard water is originally low. Not all the hardness is removed from the water at the first contact with the bed of ion-exchange resin, and the water may penetrate several inches into the bed before the hardness is in equilibrium with the resin. Since sodium is added in proportion to the hardness removed, the sodium is increased as the hardness is reduced. Thus, partway down into the resin bed, the partially softened water will be high in sodium, and so not completely softened. Thus, the capacity and leakage will worsen as the total dissolved solids increases.

This effect is minimized when the resin bed depth is increased and the flow rate per cubic foot is reduced. Larger softeners will thus produce the best performance when highly mineralized waters are to be softened.

The presence of turbidity or particulate matter in a water supply should also be considered. Even though a water softener does have some filtering ability, significant concentrations of turbidity should be removed by a separate filter before the water is softened. Filtration equipment of various types will be discussed later in this chapter.

The performance of a water softener may be considered unsatisfactory because of iron bleed when there is iron in a supply. Whereas a softener can remove some iron, its ability is limited to the type and amount of iron present. A softener alone should never be installed where the water is red or rusty when drawn or where iron bacteria are present in the supply.

When manufacturer's specifications are vague, the iron limits given in Table 9-5 are suggested.

Even when iron is considered in the selection of a water softener, the iron can cause a problem if allowed to oxidize and precipitate in the resin bed. If this occurs, the iron is no longer affected by the brining cycle. There are several ways to retard this. One can regenerate more frequently at a lower salt dosage, thus giving the iron less of a chance to oxidize; use salt with a special additive to redissolve any oxidized iron; or include a special cleaning-agent system with the softener. Any of these methods may reduce the chances of an iron-fouled resin bed. If the softener does become iron-fouled, it is relatively easy to clean. This cleaning procedure is described in detail in Chap. 11.

The last consideration when using a water softener is the type of salt to be used. There are two types of salt generally available: rock salt and purified salt. The rock salt can be extremely dirty, often containing up to 5 lb (2.25 kg) of insoluble matter per 100 lb (45 kg) of product.

Table 9-5. Water-softener Iron Limits

Bed volume, ft^3	*Clear water iron limit*
0.3	0.5
0.5	2.5
0.7	4.0
1.0	5.0
1.5	5.0

Purified salt has only trace amounts of insolubles. It is available in granulated, pellet, and block forms. It is generally advisable to use a purified salt to prevent fouling or clogging of brining mechanisms. If rock salt is used, the cleanest available brand should be selected.

A water softener is not the only way to deal with hardness problems. Where a water softener is impractical, certain polyphosphate compounds such as sodium hexametaphosphate can be used to help alleviate some hard-water problems. While such treatment in no way provides all the advantages of soft, hardness-free water, it can help curb scale formation within the hot-water system.

Phosphate compounds, will chelate, or "tie up," hardness ions and reduce their tendency to precipitate and form scale. They are usually added to the water supply by a chemical feeder at a level of 1 to 20 mg/L. There are two types available, the chemical feed pump, and a pot-type feeder.

The chemical feed pump is a positive displacement pump that injects chemical only when the well pump is running. The pot-type feeder is installed on the inlet-supply line. As water flows, a portion of it is bypassed through a canister containing slowly dissolving phosphate crystals. Although this system is not as accurate as the chemical feed pump, it is usually less expensive and simpler to maintain.

The concept of phosphate treatment is more an art than a science. Little can be said except that some threshold value, usually between 2 and 20 mg/L, can inhibit the scale formation of virtually any hard-water supply. Since no magic formula is available, the key to success in phosphate treatment is trial and error. The feeder must be adjusted and readjusted until effective treatment is achieved. It should be remembered that polyphosphate treatment *is not* equivalent to water softening. It can be used only to inhibit the formation of hardness scale and will not provide the other advantages of soft water, namely reduced soap and cleaning-aid consumption.

The preceding pages have discussed the advantages of the domestic water softener. Some caution must be used in the use of softened water. Persons on sodium-free or restricted sodium diets may be concerned about drinking softened water (or for that matter, unsoftened water that contains sodium). This topic is discussed further in Chap. 11.

In most cases, however, a water softener is an asset to a household. A family of four using water that is 20 gpg hard will spend about $20 on soap, detergent, shampoo, and other cleaning products each month. Estimates show that 70 percent of this expenditure goes for chemical water softeners included in the formulation of cleaning agents. In other words, $170/yr is spent on water softeners. This cost can be reduced up to 50 percent ($85) by using plain soap and soft water. This does not take into consideration the extended life of plumbing, linens, clothing, water heaters, and the factor of reduced energy consumption. When all these items are taken into account, a home water softener may well pay for itself in a few years.

The most important step in the purchase of a water softener is the selection of a reliable supplier. A water softener works harder than any appliance in the home—every time a faucet is turned on it goes to work. Because of this and the fact that it employs valves, timers, and other mechanisms, it requires occasional service and repair which should be provided by qualified personnel. But given proper maintenance, it will provide years of service.

IRON IN WATER

Iron-free water is easy and inexpensive to treat, but unfortunately, the earth's crust is 5 percent iron—and iron can exist in one degree or another in most ground water supplies. When present in levels exceeding only 0.3 mg/L, it can cause unsightly red or rusty staining of virtually anything it comes in contact with. Like hardness, its presence in water is facilitated when acidic rainwater dissolves various iron compounds.

Iron has the ability to change valence states. In its soluble, ferrous state (Fe^{+2}) it can cause taste and odor, and will oxidize on exposure to air, forming insoluble, stain-causing rust. This rust, or ferric iron (Fe^{+3}), can create havoc in plumbing systems, water softeners, and other water-using devices. Depending on its type and amount, it may require special treatment equipment.

When approaching iron problems, there are three basic questions one must answer. Also, see Table 9-6.

1. Is the water clear and colorless when drawn from the supply (indicating any iron present is ferrous iron)?
2. Is the water red, yellow, or rusty when drawn from the supply (indicating presence of ferric iron)?
3. Is there a gelatinous sludge present in water tanks of any kind (iron bacteria, i.e., *Crenothrix, Leptothrix*)?

Table 9-6. Types and Descriptions of Iron

Ferrous Iron (Fe^{2+})	*Ferric iron (Fe^{3+})*
Clear-water iron	Red-water iron
Dissolved iron	Oxidized iron
Soluble iron	Insoluble iron

When iron is present in the ferrous or soluble state, it readily and inevitably will oxidize to form rusty, ferric iron when exposed to air, heat, or oxidants such as chlorine.

Most iron enters a water supply when acidic rainwater dissolves one of the many variations of iron ore:

$$2H_2CO_3 + FeO \rightarrow Fe(HCO_3)_2 + H_2O$$

In this equation, rainwater (as carbonic acid) dissolves iron ore to form ferrous bicarbonate. The oxidation of dissolved colorless ferrous or manganous bicarbonate in iron- or manganese-bearing water supplies to insoluble iron or manganese rust is demonstrated below:

$$\begin{array}{ccccccc} 4Fe(HCO_3)_2 & + 2H_2O + & O_2 & \rightarrow & 4Fe(OH)_3 & + 8CO_2 \\ \text{or} & & & & \text{or} & \\ Mn(HCO_3)_2 & & & & Mn(OH)_3 & \\ \text{Soluble ferrous} & \text{Water} & \text{Oxygen} & & \text{Insoluble ferric} & \\ \text{or manganous bicarbonate} & & & & \text{or manganic hydroxide} & \\ \text{(in water)} & & & & \text{(iron or manganese rust)} & \end{array}$$

There can be many forms of iron present in water; these equations are meant only as examples.

The presence of iron bacteria can complicate matters further. The iron reactions described above cause problems that are generally aesthetic in nature. When certain bacteria are present the problems become more complex. A filamentous sludge develops which is capable of clogging valves, plumbing fittings, and water-using appliances, often rendering them useless. Even small amounts of such bacteria can be a problem.

The removal of iron can be one of the more difficult tasks in water conditioning. There are several available alternatives. One is the use of a water softener, which was mentioned earlier.

Water softeners should be applied to remove iron when the water is clear when drawn and no iron bacteria are present. Manufacturers' specifications for the upper limit for removal of iron with a softener should be complied with, no additional treatment should be followed, and the preventive measures outlined in Chap. 10 should be followed.

When the type or amount of iron exceeds the treatment limits of a water softener, additional measures will be necessary. A catalytic oxidizing filter may be useful. This type of filter employs a medium that has been impregnated with various oxides of manganese. It may act as a catalyst or, in some cases, directly participate in the oxidation reaction.

As ferrous, iron-bearing water passes through such a filter, the iron comes in contact with the medium and oxidizes to form insoluble ferric iron. The resulting rust particles are then trapped in the filter bed. As the rust builds up in the filter, it will eventually clog up and must be cleaned. This procedure, called backwashing, is accomplished by flow of water through the filter and thereby causing the rust to be shaken loose and washed down the drain. Backwashing will be required more or less frequently, depending on the size of the filter, the amount of iron in the supply, and the quantity of water used.

The purpose of the oxidizing filter is to assure oxidation and flocculation of supply, which is accomplished in one of two ways. It has been rather conclusively established that dissolved oxygen and ferrous iron are adsorbed on the catalytic surface in relatively high concentrations. The ferrous iron will be present as ferrous bicarbonate, ferrous hydroxide, or ferrous chloride; but because of the excess of carbon dioxide gas in rainwater, it will most often be present as ferrous bicarbonate. As such, the iron will have a lower activity and remain in solution even when oxygen is present. This remains true even at a pH above 7.0. However, when concentrated with dissolved oxygen on the filter surface, the iron oxidizes to form a filterable hydroxide:

$$4Fe(HCO_3)_2 + O_2 + 10H_2O \xrightarrow{\text{high pH}} 4Fe(OH)_3 + 8H_2CO_3$$

Sufficient alkalinity must be present to block this reaction and allow flocculation.

When the water supply is devoid of dissolved oxygen, the medium participates directly in the reaction. In this instance, the higher oxides of manganese release oxygen to the adsorbed iron, causing oxidation and precipitation. As with the catalytic reaction, the water should be sufficiently alkaline to permit good flocculation. In such installations, periodic regeneration of the bed with potassium permanganate is necessary to maintain an adequate supply of the manganese oxides on the coating. This procedure is outlined in Chap. 11.

Regardless of the chemical function of the oxidizing filter, there are suggested water-analysis and flow-rate standards that must be considered if the filter is to operate efficiently. (See Table 9-7.)

When all considerations are satisfied, one would expect essentially

Table 9-7. Operational Criteria for Iron Filters

Factor	*Range*
pH	7.0 or higher*
Alkalinity	two-thirds of total dissolved solids level
Service flow rate	Maximum: 5 gal/(min)(ft^2) surface area
Backwash flow rate	Minimum: 8–10 gal/(min)(ft^2) surface area

* While some manufacturers claim effectiveness at pH as low as 6.5, higher ranges promote better flocculation and filtration. Lower pH can also result in dissolution of the coating.

complete iron removal. In practice, however, this is rarely true, for the results are more commonly 75 to 90 percent removal. Since iron can cause stains at levels as low as 0.3 mg/L, an oxidizing filter alone will usually not eliminate all the problems associated with iron. It is good practice to install a water softener following an iron filter to remove the remaining iron and any hardness that may be present. The practical limits (in terms of iron content) for an iron filter are ferrous iron, 20 mg/L; ferric iron, 20 mg/L; and ferrous or ferric, bacterial iron, 5 mg/L (with periodic chlorination of the system).

If we assume a 10 to 25 percent iron bleed, we can remove the remaining 2 to 5 mg/L with a water softener. In the case of bacterial iron, even a slight bleed could foul a softener and additional treatment would be advisable where the iron level exceeds 5 mg/L.

All in all, the oxidizing filter, used properly, is one of the most effective means of controlling iron. Backwash and regeneration intervals are discussed in Chap. 11.

When the iron level exceeds 20 mg/L or when higher amounts of bacterial iron are present, additional treatment will usually be required for good results. One such method is called oxidation filtration. The iron is "pre-oxidized" and the resulting floc is removed by a filter. This is usually accomplished by injecting chlorine into the inlet-supply line ahead of a pressure or storage tank. The iron is oxidized and precipitated in the tank and removed by the filter.

Four factors must be considered when chlorination is applied: chlorine demand, contact time, chlorine residual, and alkalinity. Since the chlorine demand can fluctuate, contact time and residual chlorine are usually the controlling factors and can be regulated through system modification.

Chlorine is usually fed to the water supply with a chemical feed pump. The feeder is wired to the pressure switch so that it will inject chlorine only when the pump is running. If large enough, the pressure tank may be used for detention. A tank 10 times the size of the system flow rate (in gallons per minute) will usually provide adequate contact time.

When the pressure tank provides too little contact time superchlorination is often used. The amount of chlorine fed exceeds the actual demand. This excess ensures more efficient oxidation and acts as an indicator in control of the system.

Oxidation filtration is most simply monitored by periodically checking the chlorine residual either before or after the filter, depending on the filter medium in the equipment used. The filter may use activated carbon, an oxidizing medium, or a clarifying medium. Carbon is used since it can remove chlorine along with the iron, leaving the water odorless and tasteless. However, it is not as efficient as other media for clarifying. When high amounts of iron are present, the use of a clarifying or oxidizing medium is recommended.

When using a carbon filter, check the residual ahead of the filter; when using any other medium, check the residual after the filter. This ensures control of bacterial activity within the filter itself. In any case, the residual should be maintained at 0.5 to 1.0 mg/L free chlorine. As the iron level or bacterial activity fluctuates, the chemical feed pump will require adjustment.

When applying this system to water having a low pH, steps should be taken to raise it to 7.0 or higher. Lower pHs inhibit oxidation and/or flocculation of the iron, but this is easily overcome by feeding a caustic material such as soda ash (sodium carbonate) in conjunction with the chlorine.

Another method of pre-oxidation is to feed potassium permanganate ahead of an oxidizing filter using a chemical feed pump. It may be injected into the line either directly ahead of the filter or ahead of the pressure tank. Sufficient permanganate is fed to cause a slight pink discoloration of the water ahead of the filter, but it must be clear when it leaves the filter. In theory, this is a fail-safe system. When the permanganate is underfed, the filter picks up the slack; and when it is overfed, it serves to regenerate the filter. Because of irregular flow rate, water-usage patterns, and oxidant demand, this system is often difficult to maintain. Prolonged over- or underfeeding will cause the system to fail.

The oxidation-filtration method is widely used to control iron bacteria. There are several types of iron bacteria common to well-water supplies. While generally thought to be aerobic, species of iron bacteria have been detected in deep wells containing no measurable amount of dissolved oxygen. The various bacteria will use the iron as a metabolite in different ways. All leave the iron deposited in "sheaths" of gelatinous matter. This, in turn, eventually forms a sludge or slime growth where water is allowed to stand for periods of time, e.g., flush tanks, pressure tanks, or humidifiers.

When iron bacteria are detected, shot chlorination is recommended prior to the installation of water-conditioning equipment. This is best accomplished as follows. Dumping 1 to 4 qt (1 to 4 L) of bleach directly into the well and allowing faucets to run until a bleach or chlorine odor is noticeable throughout the system. After the system has remained idle for at least 2 h, the water is allowed to run until the bleach odor is gone. If the well is severely contaminated, it may be necessary to repeat this procedure after 2 or 3 days and then periodically throughout the year. Since most iron bacteria are aerobic, this shot-chlorination procedure may control the bacteria and eliminate the need for continuous chlorination, especially where low levels of iron are present.

All the systems for removing iron have limitations. Table 9-8 should assist the user to select the best treatment method.

Table 9-8. Selection of Iron-Removal Method

Iron level, mg/L	*Bacteria*	*Clear or red when drawn*	*pH*	*Ratio alkalinity*	*Method*
0–5	No	Clear	7.0^+	. . .	Softening
0–5	Yes	Clear	7.0^+	2:1	Shot-chlorination Oxidizing filter
5–20	No	Clear	7.0^+	2:1	Oxidizing filter
5–20	Yes	Clear or red	7.0^+	. . .	Shot-chlorination Continuous chlorination Oxidizing filters
0–20	No	Red	7.0^+	2:1	Oxidizing filter
20–30	No	Clear or red	7.0^+	. . .	Continuous chlorination, Oxidizing Clarifying filter

When extremely high iron levels are present, it may be necessary to double up on some equipment. In such extremes, treatment is elaborate, expensive, and difficult to maintain. However, iron levels of 40 to 50 mg/L have been effectively handled. Such a system may require two chemical feed pumps, one to inject soda ash and one to inject chlorine, a large pressurized detention tank, two filters installed in parallel and a second pressure tank. The system is diagrammed in Fig. 9-2.

Multiple-cartridge filters have been used in treating systems where adequate flow rates to operate a tank-type filter are not available. Prechlorination of the water is necessary to oxidize the iron. The water is then passed through several cartridge filters, usually in a common housing. Each cartridge is capable of delivering 1 to 3 gal/min (4 to 12 L/min) service flow, and no backwash is required. A bank of at least five filters is usually required. If fewer are used, replacement will be frequent and bothersome. Several styles of cartridges are available; most have a nom-

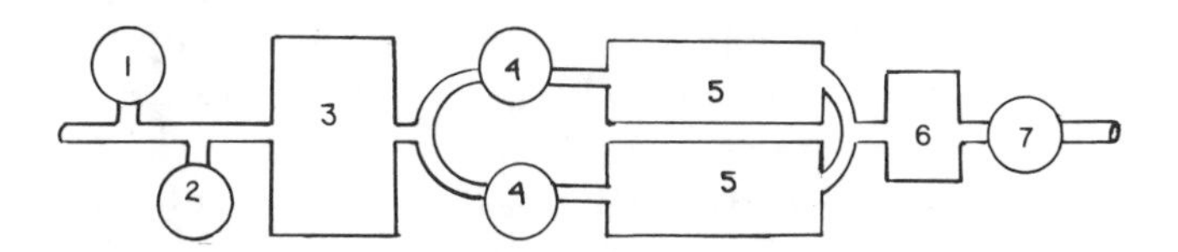

Figure 9-2. Tank-type filter system. (1) Chemical-feed pump feeding soda ash, (2) chemical-feed pump feeding chlorine, (3) 200-gal (750-L) pressurized storage tank, (4) oxidizing or clarifying filters, (5) flow-control valves (sized so that water flow through the filters doesn't exceed 5 gal/min per square foot of surface area), (6) storage tank (sized to needs of the household), and (7) water softener.

inal rating of particle size retained. Fine filters provide greater clarity, but have shorter life. Coarse filters provide less clarity, but have longer life.

A degree of iron control can also be realized by injecting controlled amounts of polyphosphates to the supply. When used in conjunction with a water softener, ferrous, nonbacterial iron at levels of up to 10 mg/L can be treated. The phosphate should be fed after the softener. The phosphate ties up or chelates the ferrous iron in much the same way as described in the section on hardness. This method is most commonly used when space or limited flow rates prohibit the use of an iron filter. Phosphate treatment will almost certainly be less than 100 percent effective.

Iron problems are primarily aesthetic in nature. Consequently, unless the iron content is maintained at 0.3 mg/L or less, the treatment must be considered inadequate. All the iron-removal methods described can be successful if all criteria are met. Great care should be taken and expert help sought when selecting the method and equipment to be used.

MANGANESE IN WATER

Manganese will also cause aesthetic problems. Gray to black staining is apparent at manganese levels as low as 0.05 mg/L. Manganese levels of up to 2.0 mg/L may be removed by a softener if measures are taken to prevent resin fouling (see Chap. 11). When the level exceeds 2.0 mg/L, oxidizing filters or oxidation-filtration methods must be used. The procedures described in the discussion of iron apply equally here. Since the problem is relatively rare and often difficult to overcome, expert help should be sought to establish the level of manganese and the proper method of removal.

CORROSIVE WATER

There are two basic treatment methods available to neutralize an acid water: tank-type neutralizing filters and chemical-feed pumps supplying a caustic material such as soda ash. The tank-type neutralizer is modeled on chemical reactions which occur in nature. Mechanically it is much like the iron filter described above, but it employs a different medium—crushed limestone, magnesia, or a mixture of both. Usually, an acid water is rain containing carbonic acid. When passed through a neutralizing filter, such water dissolves the limestone and neutralizes the acid:

$$H_2CO_3 + CaCO_3 \rightarrow Ca(HCO_3)_2$$
$$2H_2CO_3 + MgO \rightarrow Mg(HCO_3)_2 + H_2O$$

These reactions are identical to the ones described in the discussion on hard water; indeed the neutralizing filter is a kind of water hardener. One of the drawbacks of this type of neutralization is that the hardness of the water can increase up to 120 mg/L depending on its pH. Other limitations must be noted. First, the limestone used is finely ground and subsequently can cause a high pressure loss across the filter. Second, the flow rate must often be restricted to ensure adequate neutralization. This, in turn, limits the treatment's results to a pH no lower than 6.0. The effectiveness of limestone filters is often marginal. They will rarely raise the pH past 6.9 or 7.0. Magnesia filters can raise the pH higher, depending on water composition. Finally, because it employs a dissolving medium, replenishing the filter bed will be necessary.

The use of a chemical-feed pump to neutralize acid water allows more options. A much lower pH can be treated without adding hardness to the water. Corrosion protection can be simultaneously provided by adding polyphosphates to the neutralizing solution, which is usually soda ash (sodium carbonate) or caustic (sodium hydroxide). The soda ash is used to treat a pH of 4.0 to 6.8. Its reaction is a simple, efficient acid-base neutralization requiring a minimum of contact time.

$$Na_2CO_3 + H_2CO_3 \rightarrow 2NaHCO_3$$

In most cases the installed pressure and plumbing system will be adequate for this reaction. The soda ash can be injected directly into the well, thus protecting the entire plumbing system, whereas the tank-type neutralizer must be installed after the pressure tank. Typically, the solution will be mixed at a ratio of 2 to 4 lb (1 to 2 kg) per 4 gal (15.2 L) of water. It is fed at a level sufficient to raise the pH to 7.0 at the faucet farthest from the installation. Some feed-pump adjustment may be required to achieve this feed rate. The system can be augmented to provide corrosion control by the addition of ½ lb (0.25 kg) of sodium hexametaphosphate to the above solution. This will prevent nuisance corrosion resulting from the aggressive nature of the water.

When the pH falls below 4.0, caustic (sodium hydroxide) must be fed. The primary consideration here is safety. Sodium hydroxide is extremely dangerous, so rubber gloves and goggles must be worn when handling it. Also, plastic, aluminum, or galvanized containers should not be used for storage or mixing. The chemical is mixed at a ratio of 2 to 4 lb to every 4 gal of water. It should be added to the water very slowly and constantly stirred. Caustic is injected into the supply in the same way as soda ash, and phosphate may also be added at the same ratio described above. When fed, the reaction is another simple acid-base neutralization:

$$NaOH + H_2CO_3 \rightarrow NaHCO_3 + H_2O$$

Either method is relatively inexpensive and easy to maintain. Each will generally consume 100 to 150 lb (45 to 67.5 kg) of chemical per year. Maintenance primarily involves keeping the solution tank full and adjusting the feeder.

When the pH level is acidic and iron treatment is required, a soda ash or caustic feed is required. The use of tank neutralizers in addition to the iron filter all too often puts excessive strain on the pressure system and restricts flow prohibitively.

Generally water with a pH of 7.0 to 11.0 will require no treatment. If the pH is higher, the well could be polluted and its water should be carefully analyzed with particular regard to potential health hazards. The well should also be inspected. If the supply is to be treated, the pH can be lowered by feeding dilute sulfuric acid in the same manner as soda ash or caustic. Since strong acids are hazardous, extreme caution must be used in making solutions. *Always add acid to water slowly, not water to acid!* Protective gloves and goggles should be worn when making acid solutions.

Any pH problem, high or low, can result in corrosion of the plumbing system. The alternative to treatment is periodic and expensive plumbing and fixture replacement. The choice should be evident.

HYDROGEN SULFIDE IN WATER

Hydrogen sulfide gas is a by-product of bacteria. Its presence in water is indicated by an obnoxious rotten-egg taste and odor. It is also corrosive and can cause black stain on silverware and fixtures. The gas is toxic if inhaled even at relatively low concentrations for short periods of time.

The treatment of water contamination with hydrogen sulfide can be accomplished in several ways. The first is filtration with activated carbon. This method is applicable only for extremely low levels, usually not more than a few tenths of a part per million. The gas is absorbed by the carbon filter until the filter is saturated, at which time it requires rebedding. Since the carbon is not selective to the sulfide gas alone, it is impossible to predict its service life. Some carbon filters have been known to last for years, others for only weeks or even days.

The chief cycles of the unit are service and backwash. The mechanics of the tank-type carbon filter are identical to those of the iron and neutralizer filters discussed earlier. However, the function of the bed is totally different. Whereas the iron filter oxidizes iron and the neutralizer dissolves limestone, the carbon filter adsorbs the gas from the water. Activated carbon is a porous charcoal made by treating bone, petroleum products, coal, peat, or wood products at high temperatures and pressures in sulfuric acid in the absence of oxygen and the presence of an activator. The acid digests any combustible material, leaving pores in the remaining carbon. Pores are called micro- and macropores, depending on their size (see Chap. 11). The macroporous areas are available to all types of chemicals, whereas the microporous areas are available only to gases with small molecular structure. Since a water supply may contain any number and type of adsorbant contaminants, bed life of the carbon is unpredictable. Neither can the bed be regenerated when it is exhausted. The only recourse is replacement.

A second way of eliminating hydrogen sulfide involves oxidizing filters. Long a widely accepted method, its popularity has declined since the advent of the chemical-feed pump. Nonetheless, its application merits discussion. Hydrogen sulfide gas is readily oxidized into elemental sulfur when exposed to oxygen:

$$2H_2S + O_2 \rightarrow 2H_2O + 2S$$

This reaction occurs quite readily within the oxidizing filter. The chief drawback is that hydrogen sulfide has an extremely high oxidant demand—more than three times that of iron. The result is that the oxidizing filter requires very frequent treatment (daily in extreme cases) with permanganate. Because of this frequent maintenance, the method is often passed over in favor of continuous chlorination.

Probably the most efficient and easily maintained procedure for the elimination of a hydrogen sulfide problem is the oxidation-filtration method. The versatile chemical-feed pump is used to inject chlorine into the inlet-supply line ahead of the pressure tank. The chlorine oxidizes the sulfide and any precipitates are subsequently removed by a filter. The chlorine-sulfide reaction is spontaneous and does not require as much contact time as would iron or bacteria. In most cases a 40-gal detention vessel is adequate. In some cases, where no iron is present, postfiltration will not be required. A test should be performed before this procedure is eliminated, because the sulfur itself may form enough precipitate to require filtration. As in other oxidation-filtration methods, a chlorine residual of 0.5 to 1.0 ppm must be maintained to ensure adequate treatment.

SALTY, BRACKISH, OR SELTZER WATER

The presence of high levels of sulfates, chlorides, or alkalinity can cause a number of problems which fall into the same general category.

As we mentioned earlier, sulfate is most common in the form of magnesium sulfate, or Epsom salt. Commonly associated with bitter taste and laxative effect, it can be a very troublesome contaminant, especially to the livestock producer. Its removal can be achieved in three ways: dealkalization (water should be softened ahead of the dealkalizer), deionization, and reverse osmosis. Since deionization and reverse osmosis apply equally to the chloride and alkalinity problems, they will be discussed later. Dealkalization employs an apparatus much like a softener except for the type of resin used.

An ion-exchange dealkalizer resin is usually based on the same styrene–divinyl benzene polymer as softening resin. The softening resin is a cation exchanger because it has a sulfonic acid group added to the polymer. The dealkalizer resin is an anion exchanger because it has a quaternary ammonium group added to the polymer. For sulfate removal (dealkalization) the resin is put into the chloride form by treating it with salt (sodium chloride). The basic reaction is as follows (the copolymer bead will be referred to as $R \cdot C_3H_6 \cdot N^+$).

$$R \cdot C_3H_6N \cdot OH + NaCl \rightarrow R \cdot C_3H_6N \cdot Cl + NaOH$$

$$\begin{matrix}\text{Anion resin} \\ \text{(Hydroxide)}\end{matrix} + \begin{matrix}\text{Sodium} \\ \text{chloride}\end{matrix} \rightarrow \begin{matrix}\text{Anion resin} \\ \text{(Chloride)}\end{matrix} + \begin{matrix}\text{Sodium} \\ \text{hydroxide}\end{matrix}$$

Once in the chloride form, the resin is washed and prepared for use. The equipment used may be similar to a softener. The water passes through the unit where sulfates are exchanged for chloride ions, leaving the water free of sulfates.

$$MgSO_4 + 2R \cdot C_3H_6N \cdot Cl \rightarrow MgCl_2 + (R \cdot C_3H_6N)_2SO_4$$

$$\begin{matrix}\text{Magnesium} \\ \text{sulfate}\end{matrix} + \begin{matrix}\text{Chloride} \\ \text{from resin}\end{matrix} \rightarrow \begin{matrix}\text{Magnesium} \\ \text{chloride}\end{matrix} + \begin{matrix}\text{Sulfate} \\ \text{from resin}\end{matrix}$$

The chloride level of the water is a determining factor in the treatment's overall effectiveness. As a rule, the resin has a capacity of 5 to 13 kg/ft^3, depending on the regenerant level and the level of chlorides in the water. The resin is regenerated with a mixture of sodium chloride and sodium hydroxide [5 lb (2.25 kg) NaCl plus 0.5 lb (0.25 kg) NaOH per cubic foot].

This system can also be effectively used to reduce the alkalinity level of a water supply. The total alkalinity of water indicates the presence of one or more of three anion groups; carbonates, bicarbonates, and hydroxides. They can cause bad tastes or effervescence when present in sufficient quantities. The dealkalizer actually gets its name from this application:

$$Ca(HCO_3)_2 + 2R \cdot C_3H_6N \cdot Cl \rightarrow CaCl_2 + 2R \cdot C_3H_6N \cdot HOC_3$$

$$\begin{matrix}\text{Calcium} \\ \text{bicarbonate}\end{matrix} + \begin{matrix}\text{Chloride} \\ \text{from resin}\end{matrix} \rightarrow \begin{matrix}\text{Calcium} \\ \text{chloride}\end{matrix} + \begin{matrix}\text{Carbonate} \\ \text{from resin}\end{matrix}$$

Again, the chloride level is critical because of its constant regeneration effect. A dealkalizer should not be applied when the chloride level exceeds 50 percent of the total anion content of a water. A 70 to 90 percent reduction in the alkalinity and sulfate level of a supply can be expected when the dealkalizer is properly applied.

The chloride ion, as seen above, can actually interfere with anion-exchange techniques. It can be treated, however, by reverse osmosis (RO) and deionization, though not as selectively as the sulfate and alkaline groups. Chloride removal involves simultaneous removal of sulfate, alkalinity and, for that matter, total dissolved solids.

Reverse osmosis developed from research on natural osmotic phenomena. Osmosis is the process by which plants feed themselves. Within its root structure, a plant has a solution with a high concentration of dissolved solids. This solution tends to dilute itself by drawing water from the earthbound side of the root through a thin membrane that acts as the root wall. In reverse osmosis, pressure is applied to the concentrated side of a membrane, forcing the solution through it. The minerals present in the solution are ejected by the membrane or wall and only the "pure" water is able to pass through.

There are basically two types of reverse-osmosis (RO) membranes available today, the cellulose acetate (or CA) membrane, and the nylon membrane. The cellulose acetate has traditionally been available as a spiral-wound, or jelly-roll-type, membrane. The nylon counterpart is available as a hollow-fiber membrane consisting of millions of hollow nylon fibers in a bundle. Very recently the CA type has become available in the hollow-fiber style. Virtually all larger RO systems employ essentially the same components with the exception of the various membrane styles. A typical RO unit will consist of a high-pressure pump (200 to 400 lb/in^2), an RO module, and pressure gauges. See Fig. 9-3.

Small RO systems are available that do not require a high-pressure pump. Instead, they operate on the available pressure. Because of this lower operating pressure, the amount of product water delivered is reduced. These small systems usually have a capacity to produce about 5 gal of product water per day. They include a small pressure tank as a product water reservoir so that the instantaneous service draw can be of 1- or 2- gal volume at a rate of about 1 gal/min.

The pump will increase the water from line pressure (20 to 60 lb/in^2) to 200 to 400 lb/in^2 within the module. In the module, the water is

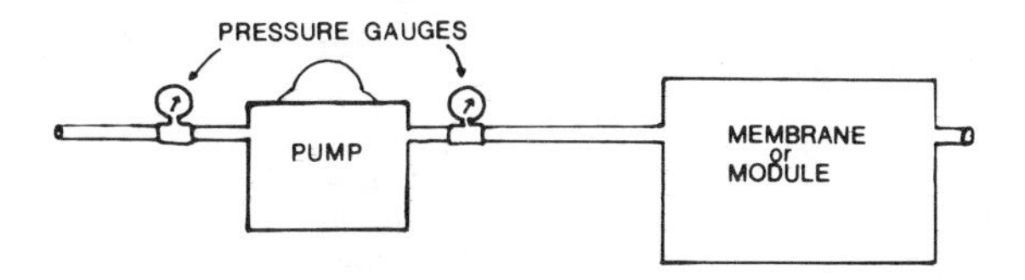

Figure 9-3. Typical setup of a reverse osmosis unit.

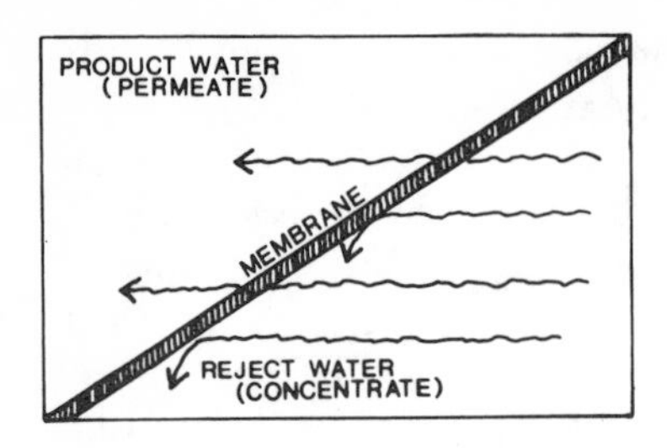

Figure 9-4. Water treatment with an osmotic membrane.

forced through the membrane wall (Fig. 9-4). This wall collects the solids and must be constantly washed by the incoming supply. This wash water is sent to the drain. The RO unit then cleans itself. This convenient feature however, is not without a drawback—namely, water is wasted when it is sent to the drain. It is common practice to dump 1 gal (4 L) of waste water for every 1 to 3 gal (4 to 12 L) of water produced. The greatest significance of this fact is its bearing on the size of the pretreatment plant.

Probably the most important factor in RO application is the proper pretreatment system. Success cannot be achieved unless a very good quality of water is supplied to the RO unit. The degree and nature of the pretreatment greatly depends on the type of membrane to be used, as Table 9-9 indicates.

The nylon membrane requires, in essence, soft, iron-free, chlorine-free, organic-free, sulfide-free, turbidity-free water. The pretreatment can be quite a task.

The CA membrane is more versatile. Whereas hardness should be removed, it can be tolerated if an acid, sulfuric acid, for example, is fed ahead of the RO unit. Such acid feed can also control the natural breakdown of the cellulose acetate and so is recommended in many cases.

Table 9-9. RO Pretreatment

Hardness	*Nylon, 17 mg/L maximum*	*CA, 17 mg/L maximum (unless acid-fed)*
Iron	0 mg/L	0 mg/L
pH	4 to 11	6.0 to 6.5
Chlorine	0.1 mg/L maximum	No limit
Organics	None	None
Turbidity	None	None
Manganese	0 mg/L	0 mg/L
Hydrogen sulfide	0 mg/L	0 mg/L
Suspended solids	0 mg/L	0 mg/L

Table 9-10. RO Reject Values of Various Salt Solutions

Salt	*Percent rejection*
Sodium chloride	98
Sodium carbonate	99
Sodium nitrate	90
Sodium sulfate	99
Ammonium nitrate	80
Calcium chloride	99
Calcium carbonate	99

The presence of free chlorine will keep the CA membrane free of bacteria, which readily feed on the cellulose acetate. Except in the cases of these pollutants, the pretreatment requirements are similar to those of the nylon membrane. In either application, it is very important to monitor the pretreatment system routinely to ensure good performance and long membrane life.

The overall functional performance of an RO system is quite impressive. It generally delivers water 90 percent free of the mineral foulants in the influent supply. The rejection rate varies for different minerals, as illustrated in Table 9-10. In practice, the relatively low rejection rate of nitrates may drop to 50 percent.

There are several factors that determine the daily productivity of the RO unit. Water temperature is by far the most important. A typical RO system will produce water much more efficiently at higher temperatures. In fact, it will produce only half the quantity at 40°F that it would produce at 80°F.

Most RO systems require a storage and repressurization system. The storage system should be sized to provide enough water for a full day's use. If the supply is only for drinking and personal use, a factor of 15 gal/(person)(day) [57 L/(person)(day)] is used. If all water is treated, the factor used is 60 gal/(person)(day) [230 L/(person)(day)]. The typical layout is shown in Fig. 9-5. The entire system is obviously very sophis-

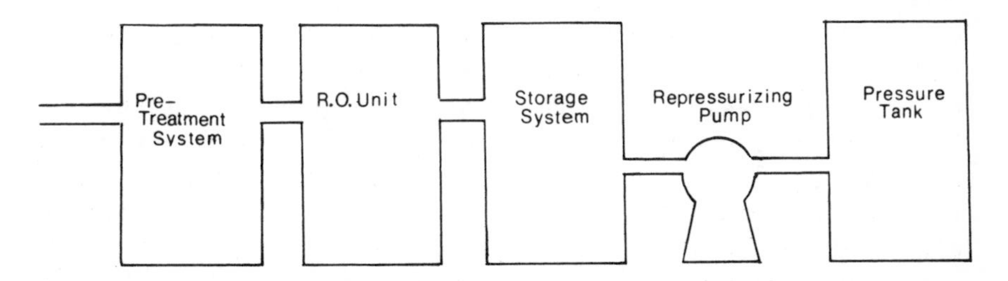

Figure 9-5. Organization of a water-supply system with a reverse osmosis unit.

ticated and requires frequent maintenance. These topics will be discussed in Chap. 11.

NITRATES IN WATER

The presence of nitrates in a water supply can cause a serious problem. Relatively low levels can cause methemoglobinemia in infants; but even high levels cause no apparent problem to an adult consumer. The source of nitrate contamination is often the matter of greatest concern. Nitrates rarely occur naturally in a water supply; their presence is usually a sign of pollution. Human, animal, and industrial waste, along with commonly used fertilizers and agricultural chemicals, are the primary sources of waterborne nitrogen. In many instances, the elimination of a nitrate problem is best accomplished by relocation or reconstruction of the well or by elimination of the source of pollution.

When nitrate removal from the water supply is the only practical alternative, the selection of a method is the first step. Whereas a great deal of research has been done in the area of nitrate removal, little of this technology is being used in the field. However, several ion-exchange processes are commonly used.

As demonstrated, we are producing sodium chloride in the effluent supply. As can be expected, the resin will not be selective to the nitrate ion. Indeed, nearly all other exchangeable ions will be affected by the process. It thus becomes important to discuss the affinity of the resin for the various anions.

In the exhaustion cycle of the resin, the ion with the lowest affinity would "break through" first and the ion with the highest affinity would appear in the effluent last. The expected order of affinity would be as follows:

$$\text{Sulfate} > \text{Nitrate} > \text{Nitrite} > \text{Chloride} > \text{Bicarbonate}$$

This means that the resin's affinity for sulfates is higher than nitrates, the affinity for nitrates is higher than nitrites, and so on. In practice, one could expect the bicarbonate ion to break through first, the sulfate ion last. Initially then, the effluent water would be relatively free of all ions except the chloride (we use the resin in the chloride form).

As the unit begins to exhaust, the ions with a higher affinity—such as sulfates and nitrates—will displace the bicarbonates, and eventually all of the bicarbonates will be displaced. At that point, the sulfate ions having a higher affinity will begin to displace the nitrates and nitrites from the bed. The effluent water will then contain increasingly higher levels of nitrates. If the cycle is continued, nitrate levels in the effluent can greatly exceed the level of nitrates in the raw water—a potentially dangerous situation. Because of this possibility, it is essential that the

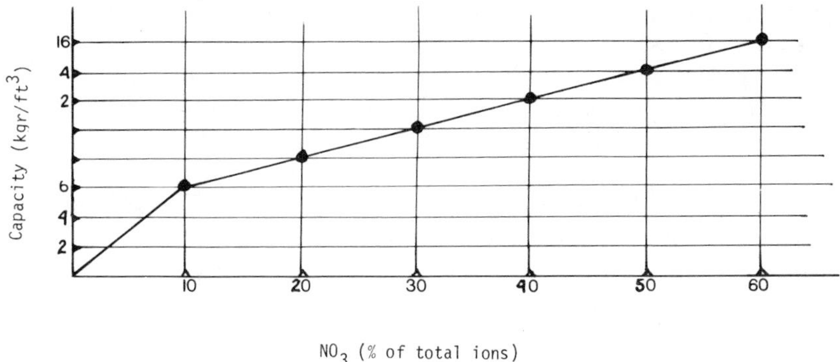

Figure 9-6. Nitrate-removal capacity (strong-base anion resin).

equipment be of proper size and so set up as to insure regeneration before the breakthrough point.

The selection of the size and regeneration level is largely dependent on the influent water quality. The primary factors are the ratios of nitrate to total anion and of nitrate to sulfate. The data in Fig. 9-6 provide some of the guidelines needed to properly predict the performance of the unit.

As shown in Fig. 9-6, the higher the nitrate content, the more efficient the unit will be. A water containing nitrate levels that are 20 percent of the total anions will reach the nitrate breakthrough point much sooner than a water with a nitrate level that is 50 percent of the total anion.

Figure 9-7 demonstrates the effect of sulfate levels on nitrate-removal efficiency.

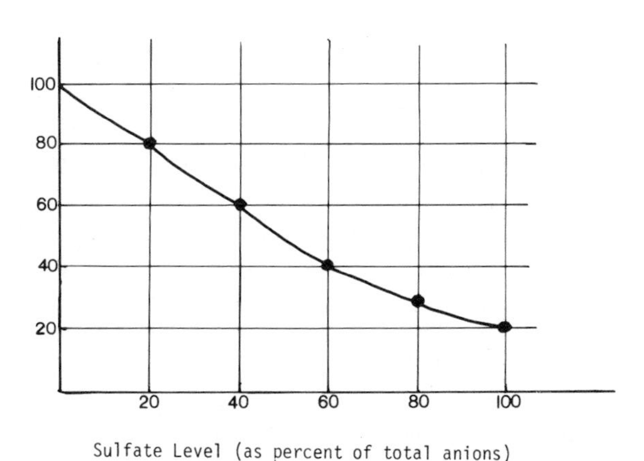

Figure 9-7. Nitrate removal—sulfate level.

Thus, as the sulfate level increases, the capacity of the resin to produce nitrate-free water decreases.

Let us look at an example of the use of this information to predict the capacity and effectiveness of a unit. The total anion content of the water we are to treat is 420 ppm, broken down as follows:

Chlorides	30 ppm as $CaCO_3$
Sulfates	130 ppm as $CaCO_3$
Bicarbonates	170 ppm as $CaCO_3$
Nitrates	90 ppm as $CaCO_3$

Nitrates as percent of total anions = 90/420= 21 percent

Sulfates as percent of total anions = 130/420= 31 percent

Using Fig. 9-6 we can determine that the capacity will be 9 kgr per cubic foot of mineral. Using Fig. 9-7 we can see that the resin will be about 70 percent efficient because of the sulfate level. We multiply 9 kgr × .7 and see our capacity for nitrate breakthrough is about 6.3 kgr/ft^3. Use of these formulas will provide safety factors for regeneration before the nitrate breakthrough.

An ion-exchange process applicable to nitrate removal is complete deionization. This process utilizes both strong acid cation-exchange and strong base anion-exchange resins. They can be housed in separate tanks or in a single-tank, mixed-bed system. When used in deionization, the cation resin is in the hydrogen form, designated as H·R. The anion resin is used in the hydroxide form and will be designated as R·OH.

The exchangeable ions, for example sodium nitrate, will be replaced by hydrogen and hydroxide ions to form water:

$$Na^+ + NO_3^- + H \cdot R \rightarrow Na \cdot R + HNO_3$$
$$H^+ + NO_3 + R \cdot OH \rightarrow H_2O + R \cdot NO_3$$

This process is extremely efficient. It can be expected to remove virtually 100 percent of all ions present in the water. When the anion resin reaches the exhaustion point, the same affinity principles discussed earlier will apply. The chief problem involves regeneration of the equipment. The resin used in the deionizer is regenerated using hydrochloric acid for the cation bed and sodium hydroxide for the anion bed. These materials are obviously extremely corrosive and hazardous. Consequently, most deionizers are either exchange tanks or throwaway cartridges. Either type would be expensive to use for nitrate removal.

Reverse osmosis is another method commonly thought of for use in

nitrate removal. Two approaches can be considered. The first is th
of a small, line-pressure module installed only for drinking and cook-
ing water. Modules capable of producing 1 to 10 gal/day of RO product water are available. The units are fairly efficient and relatively inexpensive. The second method involves use of a larger RO system to treat the entire water supply. The parameters outlined in Chap. 11 apply for such a system. The main consideration when using RO for nitrate removal is rejection rates. Whereas most field tests show a rejection rate of 80 to 90 percent, the system should be monitored frequently to ensure adequate treatment.

REFERENCES

American Association for Vocational Instructional Materials, 1973: *Planning for An Individual Water System.*

American Water Works Association, 1971: *Basic Water Treatment Operator's Manual.*

F. E. Myers and Bro. Co., 1968: *Myers Water Conditioning Technical Manual,* Ashland, Ohio.

10
Water for Special Uses

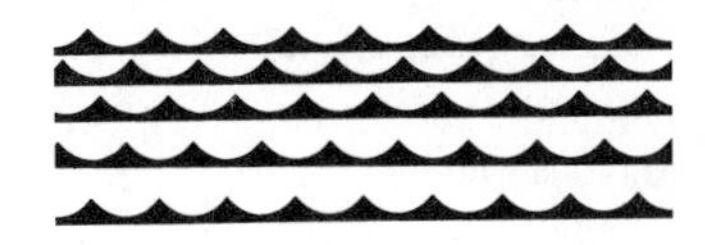

WATER FOR KIDNEY DIALYSIS

The use of kidney machines in the home has expanded rapidly and will probably continue to do so. The development of the in-home artificial kidney machine has opened a new area in domestic water treatment.

Such machines require water of a certain quality for the dialysis process. No generally accepted standards exist, but the Association for the Advancement of Medical Instrumentation (AAMI) has proposed a set of standards. Their proposal acknowledges that water used for dialysis should be of considerably higher quality than one can typically expect from water out of the tap. Table 10-1 outlines the suggested AAMI Standards.

Specific recommendations for water treatment should be left to the physician in charge, since the effects of some contaminants are not fully understood.

The most difficult step in selecting a treatment system is establishment of the desired water quality. A complete, detailed analysis of the water must first be done. A physician should then recommend which undesirable elements be removed and to what level. While only meant as a guide, not a recommendation, Table 10-2 can be used to establish the basic system.

Chapter 9 is a useful guide to the application of water-treatment equipment.

Proper sizing is obviously critical to the success of the installation. The key factor in sizing is which of two types of kidney machine is to be used: the batch machine or the single-pass machine.

A batch machine usually requires 40 to 50 gal (150 to 190 L) to complete its run. This volume must be available on demand. Since

Table 10-1. Suggested Standards for Water Used in Kidney Dialysis

Contaminant	*Maximum allowable concentration (mg/L)*
Calcium	10
Magnesium	3
Potassium	8
Sodium	150
Fluoride	0.2
Chlorine	0.5
Chloramine	0
Nitrates	2
Sulfates	100
Trace metals*	0.1

* Includes copper, mercury, zinc, tin, and barium, each limited to 0.1 mg/L.

water storage is not advisable because of potential bacterial contamination, the system must be designed to deliver the treated water in a relatively short period of time. A deionizing (DI) unit, in conjunction with any other needed equipment such as carbon filtration (CF) is usually ideal for this situation. The system should be capable of delivering 1.5 to 3 gal/min (6 to 12 L/min).

A single-pass machine demands water continuously at a relatively low flow rate, usually 7 to 10 gal/h (25 to 40 L/h). For this machine, 250 to 500 gal/day (1000 to 2000 L/day) reverse-osmosis (RO) systems work nicely. They can deliver the required flow without the need for storage.

Table 10-2. Potential Hazards and Their Removal

Contaminants	*Protection concern*	*Equipment selection**
Hardness	Patient, equipment	S
Iron/manganese	Patient, equipment	S, or IF
Total dissolved solids	Patient, equipment	RO or DI
Nitrates	Patient	RO or DI
Sulfates	Patient	RO or DI
Chlorine/chloramine/organics	Patient, equipment	CF
Sodium	Patient	RO or DI
Potassium	Patient	RO or DI
Fluorides	Patient	RO or DI
Copper	Patient	RO or DI
Aluminum	Patient	RO or DI

* RO = Reverse Osmosis; DI = Deionization; S = Softening; IF = Iron Filtration; CF = Carbon Filtration.

The water-treatment system chosen should be one selected and designed by capable water-treatment technicians with the advice of the consulting physician. The performance of the system should be monitored carefully, and, where possible, a backup or emergency system should be available.

WATER FOR AQUARIUMS

The primary problem in home aquariums is first adapting new fish to the environment of the tank, then adapting old fish to a changing tank environment. Any rapid change in water characteristics such as dissolved solids, pH, temperature, total hardness, etc., can shock the fish and result in fatalities. It is therefore necessary to adapt the fish slowly to their new environment, particularly if the change involves a completely new water supply. This is best accomplished by adjusting the supply about 20 to 25 percent per week. A total change requires about 4 to 5 weeks, usually a sufficient time to allow the fish to adapt. Needless to say, some fish are more sensitive to change, so they require a longer changeover. Allowing sufficient time for the fish to adapt also applies when an aquarium is changed from hard to softened water.

Soft water is acceptable for most home aquariums. It generally makes cleaning the tank easier because it eliminates hard-water depositions as well as some of the cloudiness that affects tank walls. When cleaning and refilling a soft-water aquarium, one should check the water to make sure that it is soft. If it is not, fish may suffer from the shock syndrome described above.

Because it is highly toxic to the fish, control of chlorine is essential to the operation of an aquarium. Chlorine can be removed by either carbon filtration or use of dechlorination drops or tablets available at most pet shops.

The last area of routine maintenance in a typical home aquarium is pH control. The ideal pH level depends largely upon the specific type of fish in the aquarium. Most do well in slightly acidic water.

The breeding of fish is a science in itself, too complicated for any commentary here. The aquarium owner should consult available literature for advice in this area.

WATER FOR PLANTS AND LAWNS

Various forms of plant life, like fish, demand their own special water quality, dependent on species. There are, however, a few general rules that can be applied.

As seems natural, simple rainwater is often best for watering plants. Generally, the higher the mineral content of the water, the less satisfactory it is for plants. As the level of dissolved solids increases, two things may occur. First, particularly if the water is high in sodium content, the level of solids in the soil will build up as the water evaporates. This "compacts" the soil, making it difficult for the plant to get water. Second, the mineral content can exceed the plant's toxicity level.

The use of softened water for household plants will not cause a problem unless the total dissolved solids exceeds 30 gpg. When using such a supply, you must water a plant until the water "drips" from the planter. This allows the water to flow through the soil, thereby helping to deter the compaction described above. But even when this is done, one must still periodically repot the plant.

Lawns should not be watered with softened water, as most grasses have a relatively low sodium-tolerance level. Since the average sprinkling wets only the top 2 to 3 in (5 to 7.5 cm) of soil, much of the water is lost almost immediately to evaporation. Successive waterings can increase the sodium level in the top soil, eventually retarding or killing the grass. Even if grass could tolerate the soft water, it would be wasteful to use soft water because of the increased salt consumption in the softener.

WATER FOR ICE CUBES

Cloudy ice cubes have long been an annoyance. They are unsightly and cause the familiar "snowflake" effect as they melt in a beverage.

Minerals in water are in solution. They are clear and colorless and will remain so until they are made insoluble for some reason. Unfortunately, freezing a solution is a good way to precipitate dissolved minerals. Most home ice cubes are made by filling a divided tray with water and placing it in the freezer. Automatic ice-makers use the same principle; the ice is simply dispensed automatically. As the water begins to freeze, the ice forms on the outside of the tray and gradually freezes inward. The minerals in the water tend to stay in solution; they must therefore migrate to the center of the tray where the solution becomes more and more concentrated. Eventually the solution becomes "supersaturated." The minerals exceed the limits of their solubility in water and precipitate. When the cube is completely frozen, it is relatively clear on the outer edge but very cloudy in the center. The higher the mineral content, the more severe the cloudiness.

When the cubes melt, they melt from the outside in, just as they froze. When the melt reaches the cloudy center, the minerals cascade to the bottom of the beverage glass. The minerals are light and fluffy and resemble snowflakes.

Two other factors affect the appearance of ice cubes: dissolved gases and the rate of freezing. Dissolved gases, such as chlorine, cause cloudiness for the same reason as minerals. They are forced to the cube's center where they concentrate. When freezing is complete, the "gas" pockets cause the cube to look cloudy, but this cloudiness does not cause the snowflake phenomenon.

The rate of freezing determines how much opportunity the minerals and gases have to form a saturated solution in the center of the cube. The slower the rate, the more opportunity there is. Rapidly frozen cubes are not nearly as hazy as are slowly frozen cubes.

The solution to the cloudy ice cube is either very rapid freezing or demineralization of the water; neither is very practical. Small deionizers or water stills can be purchased for making ice-cube water. Here the question becomes very subjective: how important is a clear ice cube?

WATER FOR HOME HUMIDIFIERS

Humidifiers essentially are water evaporators. In one type of humidifier, water is collected in a pan and is evaporated by a fan blowing across a rotating belt or fin that acts as a wick. As the water level in the pan depletes, it is refilled, manually or automatically, and the process goes on. When the water used in the humidifier is mineralized, the minerals are left in the pan as the water evaporates. Eventually they reach their saturation point in the water and begin to precipitate, leaving a scale or sludge in the pan.

When hard water is used in a humidifier, a scale will develop when the minerals precipitate. This scale often will be rock hard and difficult to remove without a descaler, such as citric acid. Often the scale will build up severely on the humidifier belts or paddles, which then must be replaced unless they are resistant to acid- or abrasive-based cleaners.

Softened water also will leave a residue in the humidifier and on the belts and paddles. Ease of cleaning is the primary difference between hard- and soft-water minerals. Usually the soft-water residue can be removed merely by rinsing the pan and belt occasionally with fresh water. Caution should be taken, however, as some manufacturers do not recommend the use of soft water in their humidifiers. Total elimination of scale or residue buildup in the humidifier would require the use of demineralized water.

WATER FOR COFFEE

What type of water is ideally suited for making coffee? This question is often debated. Some individuals maintain that hard water makes the best coffee, but many of the finest restaurants use softened or even demineralized water for their coffee. Everyone agrees that the water must contain less than 340 ppm total solids, hard or soft, to ensure good coffee. Dissolved solids in excess of 340 ppm will often produce a bitter, salty, or mineral-tasting brew; the higher the TDS, the worse the taste. Although other factors can have a bearing on the taste of the beverage, the total mineral level has the greatest impact (see Pettyjohn, 1972).

The use of soft water for coffee has, as mentioned, been debated frequently. The primary argument for hard water is that it requires less brewing time and produces a more flavorful product. It is reasoned that water containing sodium and bicarbonate (as do most artificially softened waters) causes overextraction of the coffee, which results in bitter taste. Despite the bad reports, more and more restaurants are using soft water for coffee; they are using less coffee and in many cases report a favorable customer response. It is strongly suggested that when soft water is used for coffee 25 percent less coffee should be used to make a pot. This eliminates the unusually strong or bitter taste associated with "soft" coffee. In cases of extremely high levels of hardness, bicarbonates, or other minerals, it may be desirable to use RO, demineralized, or bottled water for preparation of coffee.

WATER FOR DISHWASHING

For dishwashing, the ideal is soft, iron-free water. It will allow the detergent to do its job unhindered. There is simply no better way to clean dishes. However, the problems with dishes do not end with merely making them clean. Indeed, such headaches as spotting, etching, and tarnishing must also be dealt with.

Spotting and etching are two separate problems. Water spots are peculiar to dishes and glassware that have been dried by evaporation, usually by drip-drying on a drainboard or heat-drying in an automatic dishwasher. Since the water used to rinse the dishes usually contains minerals, it will leave mineral spotting when allowed to evaporate. This is true whether the water is hard or soft. However, soft-water spotting is nearly always less troublesome because the minerals will redissolve during the next washing. Hard-water spots are insoluble and can build up with subsequent washing. If truly spot-free dishes are desired, towel-drying is necessary. Water spotting in automatic dishwashers can be minimized by using a surface-active rinse aid and by slowing the drying cycle.

Etching of glassware is a phenomenon of the automatic dishwasher. It is actual dissolution of the glass caused by a combination of high water temperature and strongly alkaline detergents. Etching also occurs when hard-water detergents are used with soft water. The first indication of etching is a rainbow effect when glassware is held up to light. This usually prompts the user to increase the amount of detergent because the glass does not appear clean, a move that inevitably compounds the problem. Etched glassware will eventually have a cloudy or milky appearance that strongly resembles water spotting. Unfortunately, etching is irreversible and cannot be washed away. Prevention of etching is very difficult, especially where fine china or glassware is involved. The use of properly formulated detergents can deter or minimize etching, but its prevention cannot be guaranteed. Where high-quality glassware is concerned, washing by hand is the best way to prevent etching.

Common metalware problems are tarnishing and staining. Silverware is sensitive to the chlorine contained in some detergents. This chlorine can cause the silver to turn violet or black. Such tarnishing can be prevented by using a chlorine-free detergent. If silver-plated ware is scratched or worn, a copper or green color may develop. This is caused by corrosion of the base metal where the plating is worn or flawed. Little can be done except to replace the utensils. Valuable pieces can, of course, be resilvered.

WATER FOR STEAM IRONS

The steam iron is another household appliance that will function best when demineralized water is used. The steam iron heats water in a reservoir to the point of evaporation. As the water evaporates, it leaves any minerals it contains behind to clog up the reservoir and steam vents. Whether the minerals are hard or soft, they will foul up the steam iron. The only solution is to use water free of dissolved minerals such as deionized, distilled, or RO water.

WATER FOR LAUNDRY

The use of iron-free, soft water is the best way to eliminate water-related laundry problems. Softened water will eliminate soap deposition on fabrics and eliminate the need for fabric softeners, which are nothing more than water softeners. When soft water is used, much less soap or detergent should be used: 2 or 3 oz (50 or 75 g) is usually enough.

The best advice one can give for laundering is the use of soft water and plain soap rather than synthetic detergents. Soaps are cheaper and much more efficient at removing and suspending soil than are detergents. In addition, ordinary soaps are milder than the synthetic detergents that are used in their counterparts.

It is equally important to use good laundry techniques. Automatic washers should be loaded slightly below their rated weight capacity, especially when washing white clothes. The use of the proper amount of soap or detergent will also go a long way toward satisfactory results. Heavily soiled clothes should be washed in a much lighter load to prevent redeposition of the soil on the fabric.

In summary, good soft water and soap are the shortest route to a satisfactorily clean wash. The primary elements of concern in the water are

- Calcium—Soap precipitation
- Magnesium—Soap precipitation
- Iron—Red-orange staining
- Manganese—Black staining
- Copper—Blue staining

These contaminants should be removed from the water supply before laundering. If iron or manganese stains are experienced, they can be removed by using a small amount of sodium-hydrosulfite-based resin-bed cleaner. Sprinkle a small amount of the powder over the stained area, wet, and rub gently. Then rinse the powder and stain away.

WATER AND WATER HEATERS

The advantages of using softened water in water heaters were discussed in Chap. 9. The fuel and maintenance savings are significant enough to justify the use of a water softener where hardness levels exceed 5 to 7 gpg. Softened water eliminates scale buildup in the heater and hot-water plumbing. Of course, other problems can and do exist in water heaters.

Studies have shown that water becomes increasingly corrosive when its temperature is increased. The rate of corrosion on an unprotected steel surface may increase as much as 100 percent for every 20°F increase in water temperature. Since galvanized steel is widely used for unlined water-heater tanks and hot-water plumbing, this is an obvious area for concern. Many modern water heaters employ a sacrificial anode, usually made of magnesium, to protect water-heater tanks. These anodes are effective but not in areas where the total dissolved solids (TDS) of the water is below 50 mg/L. Where low TDS does occur, the use of a zinc anode is suggested. In all water heaters, corrosion can be minimized by keeping the water temperature of 140°F or below.

The use of sacrificial anodes may lead to other problems. Where TDS levels exceed 400 mg/L, the anode rapidly dissolves resulting in a hardness increase in the hot-water supply and generation of dissolved gases in the water heater. The installation of a resistor between the tank and anode will minimize but not eliminate this potential. In other cases, the hot water generates a strong hydrogen sulfide (rotten-egg) odor where magnesium anodes are employed. Although the chemistry of this phenomenon is not thoroughly understood, the reaction apparently involves the anode, dissolved sulfates in the water supply, and sulfate-reducing bacteria. The only sure way to eliminate this problem is to remove the anode. Unfortunately, the tank is then exposed to corrosion and in most cases the warranty is voided. Periodic chlorination of the water system or periodic replacement of the anode have proven only marginally successful.

WATER AND DIET

The level of sodium in a drinking-water supply may be of concern to the consumer on a sodium-restricted diet. It is important that the consulting physician take the nature of the water supply into consideration when prescribing such a diet. Because most ground water supplies will contain some sodium, a comprehensive water analysis is needed.

The level of sodium (as Na) in milligrams per liter in a supply can be directly converted to milligrams of sodium intake per liter of water consumed. For example, if the supply contains 100 mg/L of sodium, the patient who consumes 1.5 L of water per day has added 150 mg of sodium to daily intake. The consumer who is on a strict "salt-free" diet (usually 500 to 1000 mg of sodium per day) should not drink the water. Distilled, demineralized, or bottled water would be the ideal solution.

A second consideration would be whether or not the supply is softened by ion exchange. As we discussed earlier, hardness minerals are exchanged in a water softener for sodium. Obviously, this would increase the sodium content of the water. The formula below provides a quick method of predicting the sodium added by ion-exchange softening.

$$\text{Hardness (340 mg/L as } CaCO_3) \times 0.46 = 156 \text{ mg/L sodium added by the softener}$$

Citing our example, 156 mg sodium would be added to the consumer's diet for each liter of water consumed. Some homeowners install special hard-water taps in their kitchens; water from these taps is used for drinking and culinary purposes.

REFERENCE

Pettyjohn, W. A., 1972: *Good Coffee Water Needs Body in Water Quality in a Stressed Environment*, Burgess Publishing Co., Minneapolis, pp. 194–199.

11

Care and Maintenance of Equipment

This chapter is devoted to the care of various water-treatment devices. Because of wide variations in manufacturers' designs, it is obviously impossible to discuss the specifics of particular brands or styles. Thus, the discussion of flow patterns and maintenance techniques will be of a general nature. It is hoped that this how-it-works section will provide a basic understanding of the various kinds of equipment available.

WATER SOFTENERS

Previous chapters stated that a water softener consists of three functional components: the resin or mineral tank, the brine or salt-storage tank, and the controller. Let's look at each component and its job in more detail.

The resin tank is usually constructed of fiberglass or galvanized steel and holds the ion-exchange resin that was described in Chap. 9. Because of their noncorrosive properties, fiberglass tanks have gained widespread popularity. Many manufacturers use fiberglass tanks exclusively and cover them with generous warranties. Some steel tanks are made to include a separate plastic liner or coating. Metal tanks used for water softeners should have proper protection against internal corrosion, which is normally otherwise severe in the presence of the ion-exchange resin. A resin tank for a water softener must be capable of withstanding high water pressure. Most are tested at 300 lb/in^2 and carry a continuous rated operating pressure of 100 to 125 lb/in^2.

Distributors and riser pipes are used in the tank to direct the flow of water through the unit during its various cycles. Figure 11-1 shows a typical water-softener tank and its components.

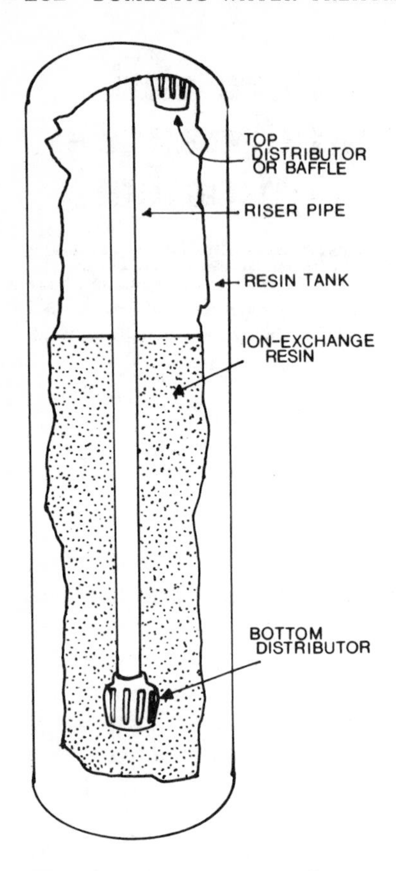

Figure 11-1. Water-softener resin tank.

Water enters the resin tank through the top distributor or baffle, which disperses the water over the top of the resin bed. This ensures uniform flow through the resin and prevents "channeling," which occurs when water is allowed to flow directly into the resin without any deflection. It results in hard-water bleed through the unit. Top distributors and baffles are usually slotted, as shown in Fig. 11-2.

The use of a distributor lessens the chance of resin loss during backwash of the softener, because the slots are sized to be just smaller than the resin beads. Baffles are popular where the softener is installed on water systems with iron-bearing water and backwashing is more critical.

Once through the distributor or baffle, the water passes through the resin bed where the softening process takes place. It is then collected in the bottom distributor and flows upward through the riser pipe and into the service lines (Fig. 11-3).

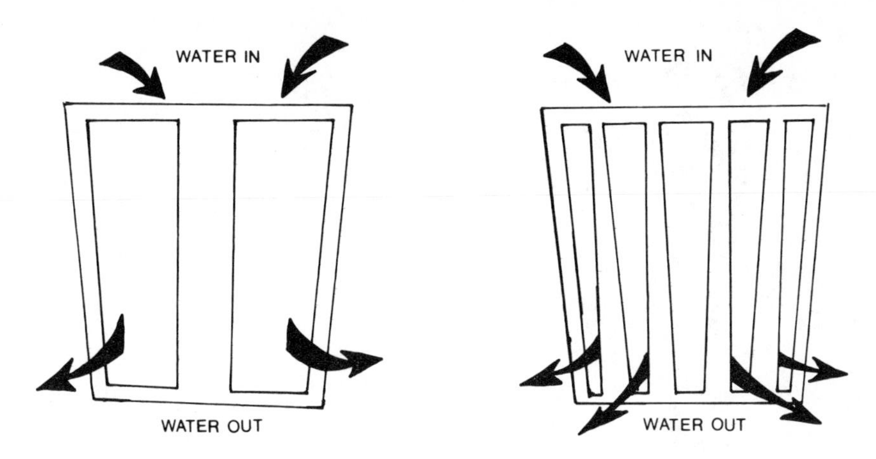

Figure 11-2. Baffle (left) and distributor (right).

Figure 11-4 puts the function of the components in perspective. The mineral tank is the workhorse of the softener, but it is rarely the cause of any service problems other than plugging.

The brine, or salt-storage, tank is constructed of fiberglass or a plastic such as polyethylene. Its primary components are a brine well, a brine valve or tube, and a salt-support system. The main job of the brine tank is to store dry salt and make up salt solutions between regenerations (Fig. 11-5).

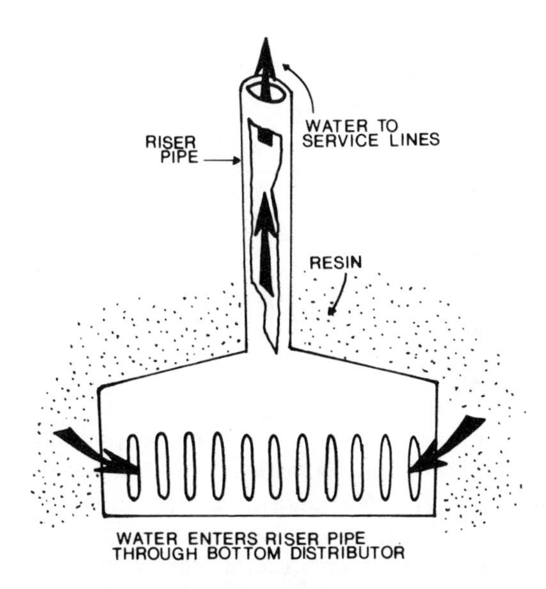

Figure 11-3. Flow through bottom distributor and riser pipe.

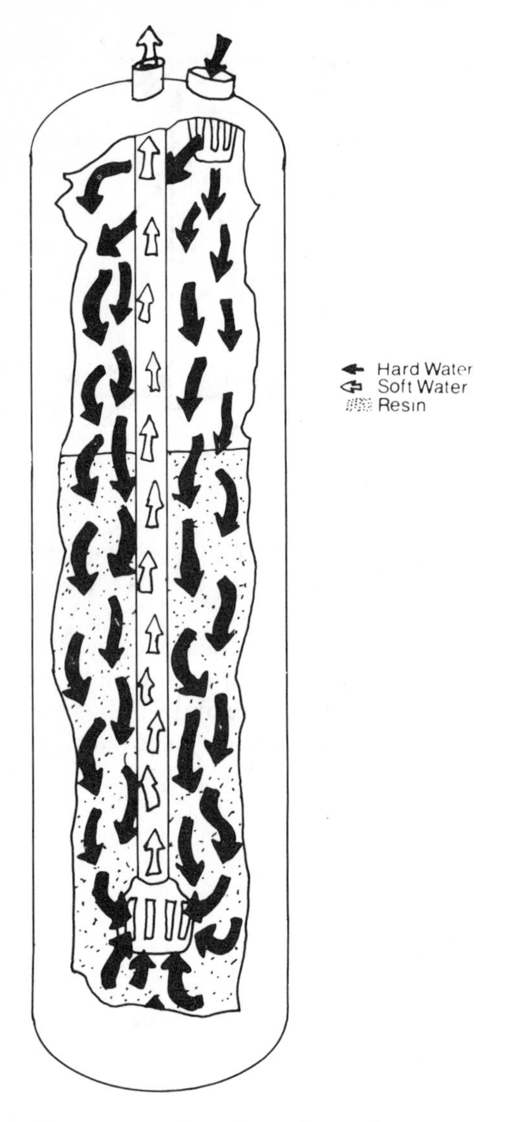

Figure 11-4. Flow through a softener.

The rate at which brine is withdrawn from a brine tank is normally independent of that system. Rather, the rate is controlled by the control valve on the softener. During regeneration, the brine is normally withdrawn to a minimum level beyond which no brine is available. The regeneration salt dosage is normally established either by using a float shutoff in the brine tank or by providing a constant refill rate for a controlled period of time. To repeat, the salt dosage is established by the amount of water that is returned to the brine tank after regeneration. This water volume should be reasonably controlled. One of the methods above is used; if both are provided, the redundancy provides a safety factor to prevent overflow if the primary refill control fails.

In the example shown in Fig. 11-5, the brine is made up below the support platform and drawn to the softener by an aspirator housed either in the brine valve assembly or in the main valve of the softener. As the level of brine solution in the tank recedes, the float drops and eventually seals off the bottom of the brine valve assembly when all of the usable brine has been drawn. The tank is refilled when the flow of water is reversed through the riser pipe. The float rises with the brine level and seals the riser in the upper position when sufficient water has been added to the tank. The water level in the tank after refill should be at least ½ to 1 in (1.25 to 2.5 cm) above the support grid (Figs. 11-6 and 11-7). In all cases, refer to the manufacturers' recommendations in the owner's manuals.

In cases where the float valve controls the amount of brine draw and freshwater refill, there is constant water pressure on the brine line while the softener is in the service cycle. If the float valve fails to seal in the up position, water will continue to flow into the tank and eventually overflow. Well-designed brine tanks are equipped with an overflow-hose adapter (Fig. 11-5). The overflow hose should go to a drain to minimize problems caused by an overfilled brine tank. Some softeners employ a float valve and a timed refill. The possibility of overflow is minimized in

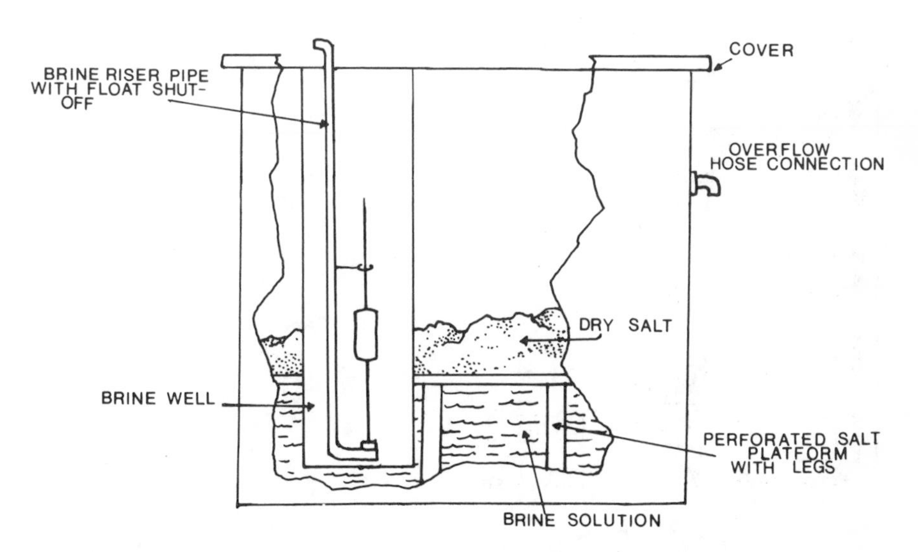

Figure 11-5. Brine tank with brine valve and salt platform.

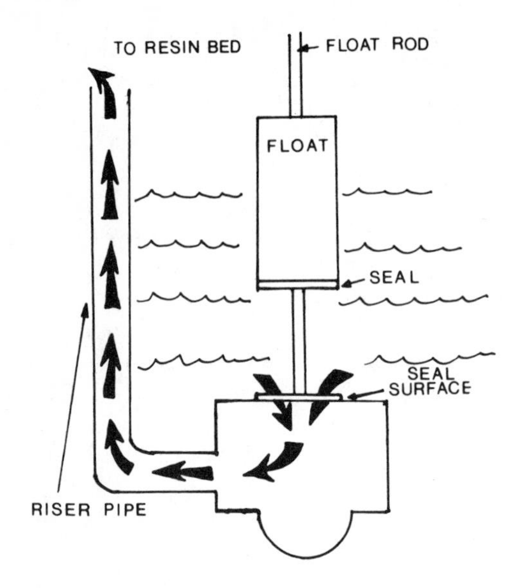

Figure 11-6. Brine valve during regeneration.

such equipment. However, overflow could occur if both the timer or timed mechanism and the float valve failed, or if the unit failed to draw brine and the float valve failed to seal during refill. Either of these cases can occur, but they rarely do.

Another method of controlling brine draw and refill is the use of a

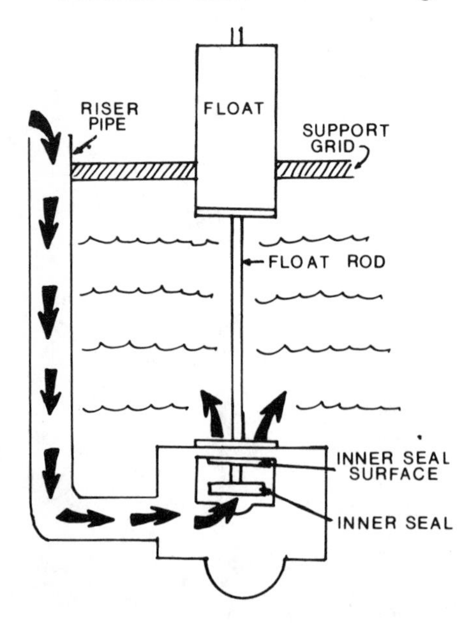

Figure 11-7. Brine valve during refill.

brine tube and timed cycles. The brine-tank assembly is roughly the same, except that a simple brine tube is located in the brine well in place of the valve system shown in Fig. 11-5. The timer controls the duration, and a flow restrictor controls the volume of water during refill. When the softener goes into regeneration, water is routed through the aspirator, which draws the brine. When the timer signals for refill, water is routed through a flow-control valve to refill the brine tank. If the refill cycle calls for the addition of 20 gal (75 L) of fresh water to the brine tank in 4 min, a 5 gal/min (19 L/min) flow control is required. If the cycle calls for a 10-gal (38 L) refill in 5 min, a 2 gal/min (8 L/min) flow control is used. In such equipment, timer failure during refill can result in overflow.

The aspirator is the backbone of the brining system. It may be housed in the main valve of the softener or in the brine-valve assembly, depending upon the manufacturer. In either case, its function remains the same. The aspirator shown in Fig. 11-8 consists of a nozzle and a venturi or throat. Fresh water passes through port *A* and creates a jet stream as it passes through the nozzle. As the jet stream passes through the venturi, it creates a suction that draws brine from the brine tank *B*. The brine is drawn through the venturi and is mixed in the chamber above the venturi. The mixed brine passes through port *C* to the resin bed of

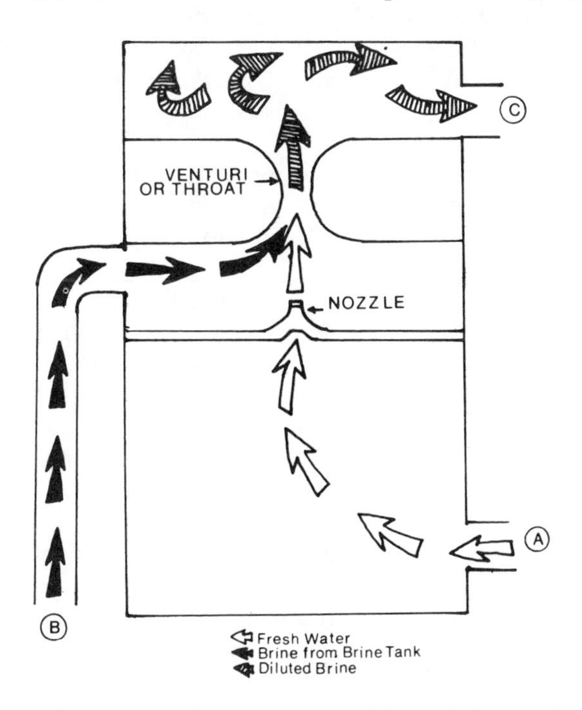

Figure 11-8. Aspirator assembly and flow pattern.

the softener where regeneration occurs. When the regeneration is float-controlled, the float shuts off and only fresh water passes through the venturi. This fresh water is used to rinse the softener. When the cycle is timer-controlled, water to the assembly is shut off, allowing no more brine draw, and the softener is rinsed.

The controller can be referred to as the brain of the softener. It consists of the main valve and the cycle control, most commonly a timer. There are many styles of water-softener valves available. They may be hydraulically operated by diaphragms or electrically operated by solenoids, motor valves, or timer-operated cams. It would be futile to try to discuss them all, so the various cycles of operation, most of which are common to all makes, will be considered. The primary job of the controller is to tell the softener what to do. A meter or a special probe (sensor) or combinations of these devices are used to signal the softener when to activate each cycle (backwash, brine draw, rinse, etc.). The main valve directs the water wherever it is needed at any time. The following diagrams show the function of the valve during each cycle (Figs. 11-9 to 11-12).

During the service cycle (Fig. 11-9), the main valve is in a neutral position. It allows water to enter through the hard-water inlet and to pass through the top distributor to the resin bed where softening takes place, through the bottom distributor, up the riser pipe, and out into the service lines of the home. When the regeneration control signals the unit to cycle, the main valve sends the softener through a number of steps. Typically the order is (1) backwash, (2) brine draw, (3) rinse and refill, and (4) back into service.

During the backwash, water flow is reversed through the mineral tank. Water is not allowed to enter the mineral tank through the top distributor. Instead, it is rerouted so that it enters the mineral tank through the riser pipe and bottom distributor. The reversed flow uplifts the bed, causing turbulence that cleans the resin of particles, washing them down the drain as the water leaves the mineral tank through the top distributor and then flows out the drain line. Hard water is available to the service lines during backwash (Fig. 11-10).

Regeneration occurs when water is routed through the nozzle and venturi and brine is drawn from the brine tank (Fig. 11-11). The brine passes through the resin bed, regenerating the resin, through the bottom distributor, up the riser pipe, and out to the drain. Hard water is permitted to bypass the softener during regeneration.

In the rinse cycle, water is still available in the service lines. Also, water is allowed to flow to the brine tank for refilling while the resin is being rinsed of residual brine. When this cycle is complete, the valve returns to the service position, as indicated in Fig. 11-12.

As mentioned earlier, the softener is directed by any one of a number

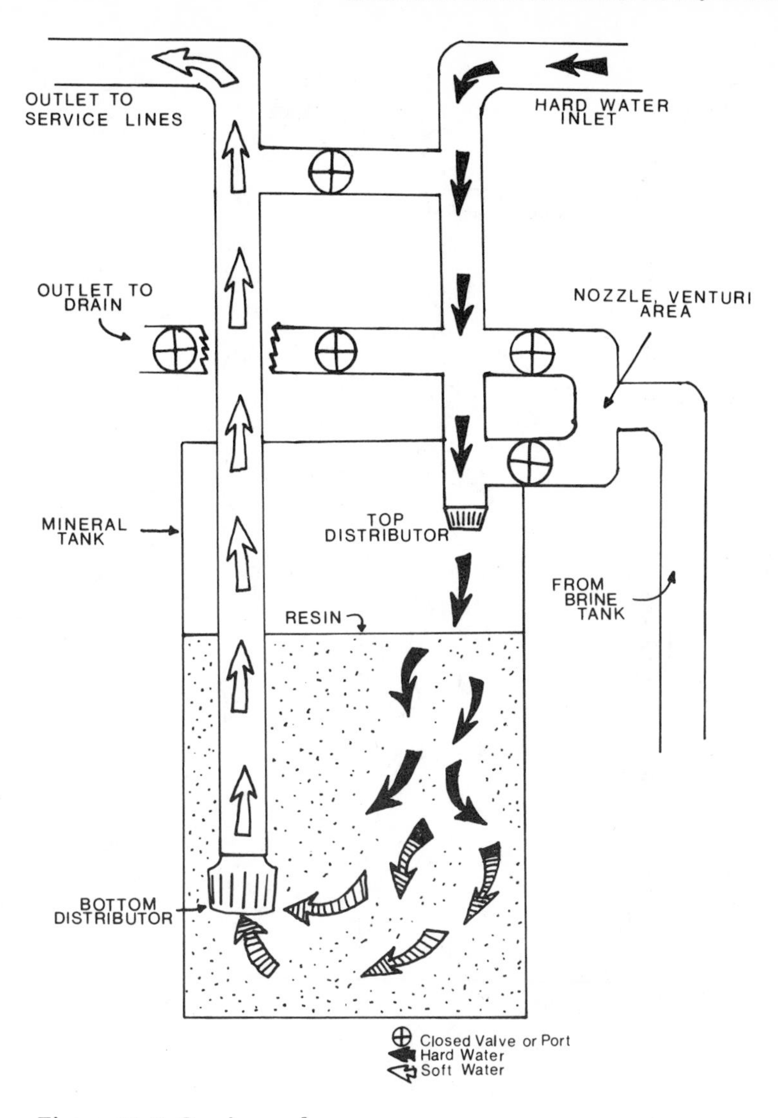

Figure 11-9. Service cycle.

of control devices, most commonly a timer. The timer is used to activate motor valves, solenoids, pistons, or cams that in turn send the valve through its various cycles. Timers are as varied as the valves they control. There are or have been 6-, 7-, 12-, and 14-day timers available. Their key function is to keep track of time and signal the unit to regenerate at given intervals. They are nearly always electrically operated and generally are the most trouble-free of the initiation mechanisms available.

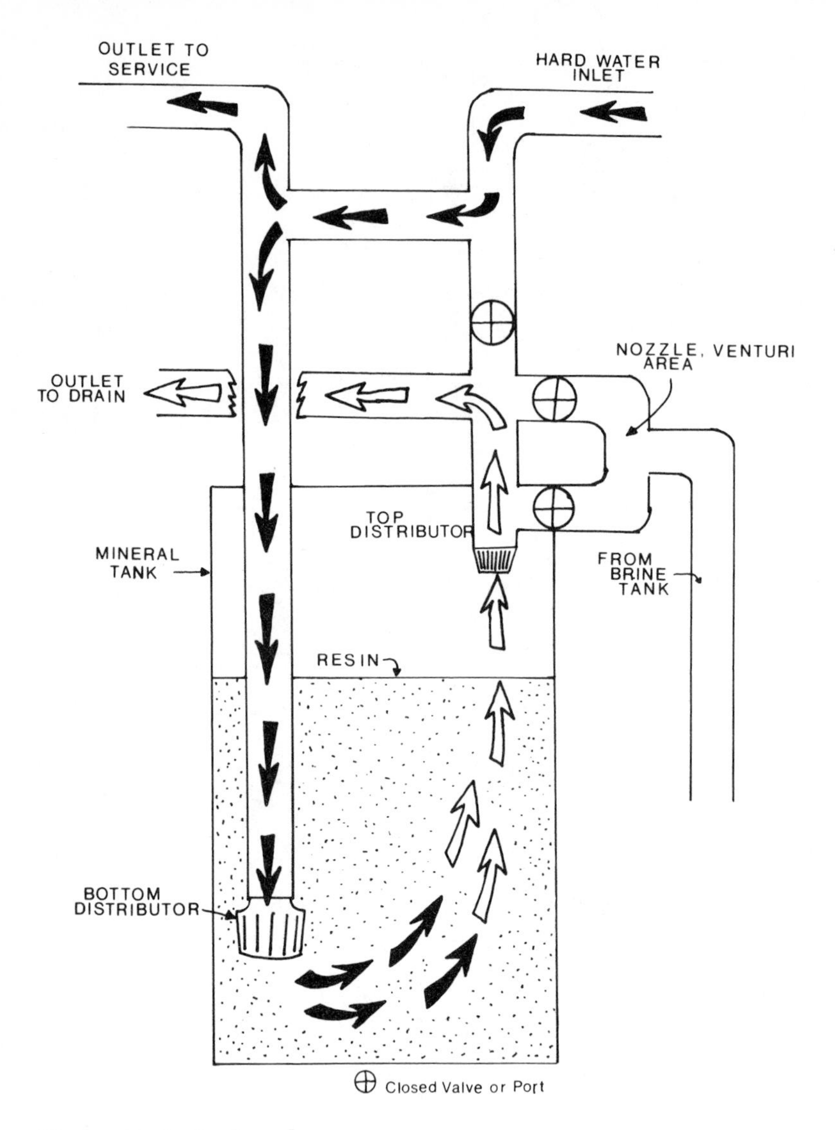

Figure 11-10. Backwash.

A second method of initiating regeneration is by use of special probes or sensors. These devices measure the change from soft to hard water, as the resin bed becomes exhausted, and electronically signal a timer to commence regeneration. Since they call for regeneration only upon demand, they have a particular advantage where water hardness or usage patterns fluctuate. However, such devices should be used with caution on high–iron-bearing waters or any water that might foul or otherwise inhibit the efficiency of the sensor.

A water meter may also be used to initiate regeneration. Again, the chief advantage is regeneration on demand. This control does not compensate for variable water hardness. Some softeners use meters that automatically regenerate the softener without the use of electricity.

The installation of a water softener is a relatively simple procedure. Most manufacturers include an installation manual with each unit. If the softener is to be installed by the homeowner, the following steps, in conjunction with the installation manual, can be followed:

1. All installation procedures must conform to local plumbing, electrical, and sanitation codes. *Drain lines from water-conditioning equipment are never connected di-*

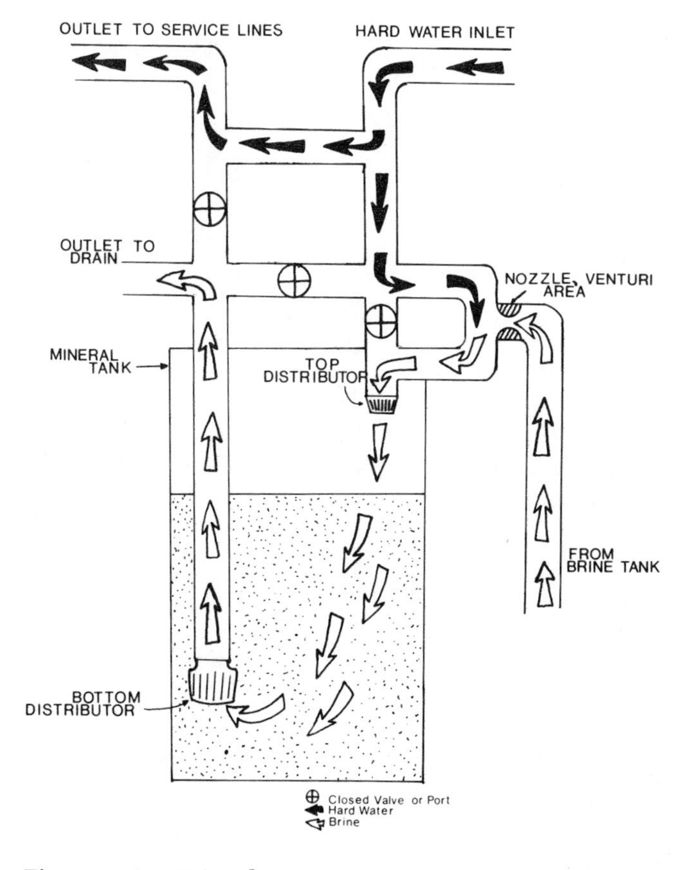

Figure 11-11. Brine draw.

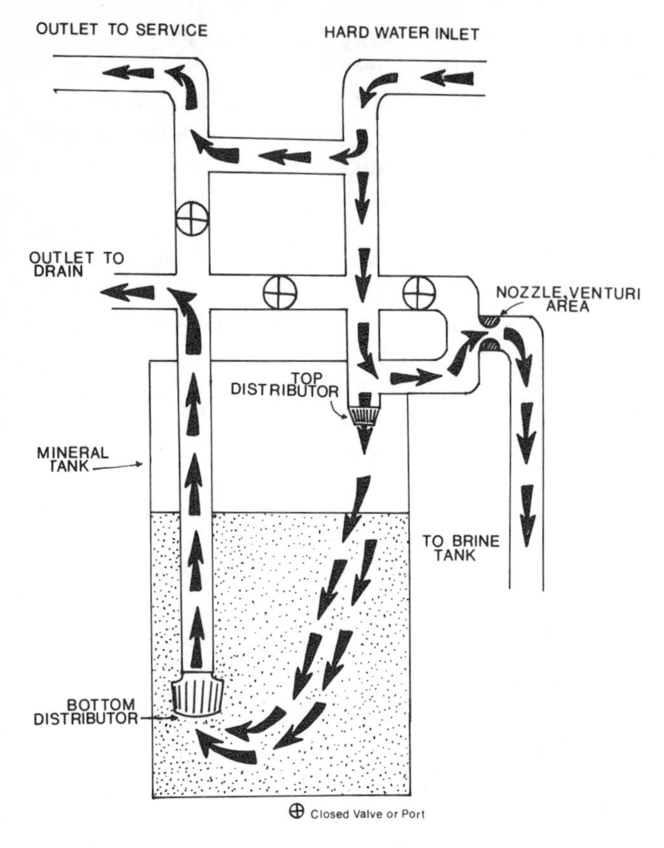

Figure 11-12. Rinse cycle with refill.

rectly to sewer lines unless a siphon break is used. Outside faucets should be plumbed to hard-water lines ahead of the softener.

2. Close main water-supply line and shut off fuel supply to the water heater.
3. Open a hot- and cold-water faucet to relieve pressure and open the valve nearest the pump or water meter to drain the system.
4. Move conditioner into position for installation. If the floor is not level, use shims to provide a solid, nonrocking base.
5. Loosely attach any flare tubes, nuts, and installation fittings to the softener.

6. Measure, cut, and assemble pipes and fittings necessary to connect unit to main supply line. Bypass valves should be included. Many dealers or outlets provide simple, easily installed bypass valves. Figure 11-13 illustrates proper installation of a three-valve bypass.
7. Be sure the hard-water inlet is aligned with the inlet of the softener.
8. Disconnect the softener from nuts and fittings and solder fittings as required. *To prevent damage to nonmetallic components, do not solder fittings while the softener is connected to plumbing.*
9. After soldering all necessary fittings, insert washers, screens, and any other accessories and reconnect softener.
10. Install drain hose on softener and place over drain.
11. Install brine-tank overflow fitting and hose and place over drain. *Do not tee overflow to softener drain.*
12. Close valves *A* and C (Fig. 11-13) and slowly open valve *B*. Check for leaks. If any resoldering is necessary, be sure to disconnect softener.

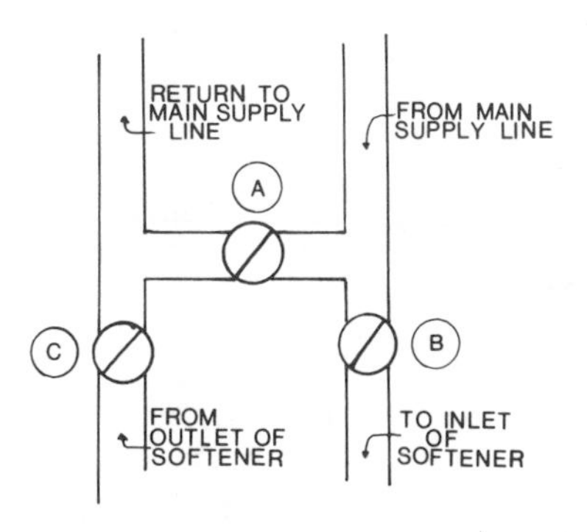

Figure 11-13. Installation of 3-valve bypass. During normal service, valve *A* would be closed and valves *B* and *C* would be open. To place unit in bypass, open valve *A* and close valves *B* and *C*.

13. Fill brine tank and adjust flood level according to specifications.
14. Add salt. Place softener into service.
15. Relight water-heater pilot.

Obviously any device such as a water softener will occasionally malfunction. Table 11-1 lists some of the more common problems, their possible causes, and suggested remedies.

Table 11-1. Trouble-shooting Guide

Failure	*Probable cause*	*Remedy*
1. Softener will not automatically regenerate	a. Timer, meter, or sensor inoperative b. Defective wiring between timer and sensor or timer and motor c. Defective power cord d. Softener is plugged into intermittent power source (i.e., socket on light switch)	a. Replace or repair defective part b. Remove connections c. Replace cord d. Connect to constant-power source
2. Conditioner regenerates at wrong time of day	a. Timer improperly set	a. Reset timer
3. Water runs to drain from softener during service cycle	a. Defective drain valve	a. Repair or replace drain valve
4. Water runs to drain from brine-tank overflow	a. If float valve is used, obstructed, or makes faulty lower seal b. Cracked or defective brine riser pipe c. Defective timer or seal within valve	a. Remove obstruction or replace seal b. Replace brine riser pipe c. Repair or replace timer or seal
5. Softener does not draw brine	a. Softener drain hose kinked or plugged, causing back pressure and killing vacuum b. Softener drain hose elevated too high	a. Remove obstruction or replace hose b. See manufacturer's height specifications

Table 11-1 (*Continued*)

Failure	*Probable cause*	*Remedy*
	causing back pressure	
	c. Brine line and/or fittings plugged or obstructed	*c.* Remove obstruction, replace as necessary
	d. Aspirator plugged or defective	*d.* Repair or replace
	e. Restriction in brine riser pipe or brine tube	*e.* Clean or replace
6. Hard-water bleed or bypass during service cycle	*a.* Manual bypass valves open or defective	*a.* Close, repair, or replace valve *A*. (See Fig. 11-13)
	b. Defective bypass seal or seat in main valve	*b.* Clean, repair, or replace as necessary
	c. Internal riser pipe cracked or not properly secured	*c.* Repair or replace as necessary
7. Salt in lines after regeneration	*a.* Low water pressure throughout system	*a.* Adjust pressure to minimum recommended by manufacturer
	b. Restricted or plugged backwash or rinse parts	*b.* Clean parts
	c. Top distributor plugged	*c.* Clean or replace
	d. Softener drain valve or hose plugged or restricted	*d.* Clean or replace as necessary
8. Using too much salt	*a.* Improperly adjusted brine valve or tube	*a.* Adjust to specification
	b. If float is used, defective float seal	*b.* Clean or replace
	c. Water leaks in brine tank causing improper fill.	*c.* See 5*a*, *b*, *c*.
	d. Timer improperly set	*d.* Reset timer
9. Air in hose	*a.* When brine float used, defective power seal	*a.* Repair or replace
	b. If timed regeneration, improper timer setting or defective seal in main valve	*b.* Repair or replace as necessary
	c. Defective air-check valve on pressure tank or air in water supply	*c.* Investigate and make necessary corrections.

Table 11-1 (*Continued*)

Failure	*Probable cause*	*Remedy*
10. Low water pressure	*a.* Municipal pressure low, low pump pressure, or defective well pump *b.* Restriction in water lines or other water using equipment *c.* Distributors and/or riser pipe plugged *d.* Plugged resin bed	*a.* Investigate and correct as necessary *b.* Investigate and correct *c.* Clean or replace *d.* Clean resin bed (See 11)
11. Iron-fouled resin bed		Clean as follows: *a.* Mix resin cleaner according to instructions on container *b.* Add resin cleaner to brine well of softener *c.* Manually initiate regeneration of softener *d.* Repeat if necessary *e.* If problem reoccurs, see Chap. 9, iron in water

CARE OF FILTRATION EQUIPMENT

The mechanics of a filter will be the same whether the filter is to be used for iron removal, taste and odor removal, pH correction, or clarification. Water filters are composed of a mineral tank and a valve. The valve is either controlled by a timer or is operated manually. As in a softener, the water enters the mineral tank through the top distributor or baffle, flows through the mineral bed, is collected in a bottom distributor, and flows up the riser pipe to the service lines (Fig. 11-14).

Of the factors that control the effectiveness of filtration, flow rate is probably the most important. There are two flow rates to be concerned about—service flow rate and backwash flow rate. The service flow rate should be restricted to 5 gal/min (19 L/min) per square foot of bed area if the undesirable contaminant is to be effectively removed. For effective cleaning of the filter during backwash, a flow rate of at least 8 to 10 gal/min (30 to 38 L/min) per square foot of bed area is required. Table 11-2 relates these requirements to various tank sizes.

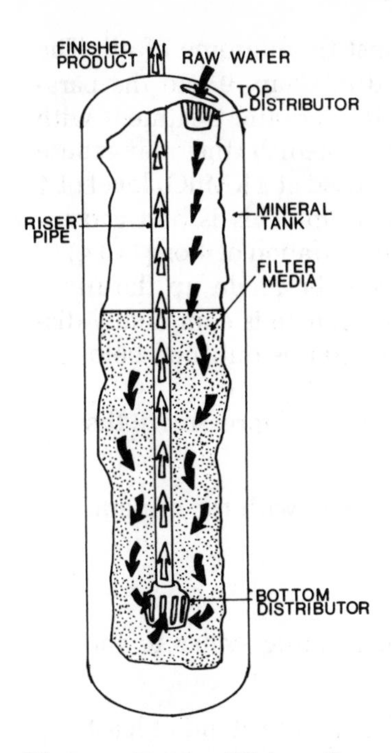

Figure 11-14. Water flow through a filter.

When selecting a filter, one should be certain that the pump and pressure system are capable of delivering the required backwash flow rate. This is particularly true when iron removal or clarification is the goal. In installations where a softener is also used, some efficiency may be sacrificed by exceeding the recommended flow rates during the service cycle. Under no circumstances should the units be insufficiently backwashed. Different filter applications require different maintenance

Table 11-2. Recommended Filter Flow Rates

Tank diameter, (in)	*Maximum service flow, gal/min*	*Minimum backwash flow, gal/min*
8	1.5	2.2–2.7
9	2.2	3.5–4.4
10	2.8	4.5–5.5
12	4.0	6.0–8.0
13	4.6	7.5–9.2
14	5.4	8.6–10.7

procedures. Iron filters are probably the most troublesome of all. The chemistry of oxidizing filters was discussed in Chap. 9 and the parameters suggested there should be adhered to. Periodic treatment with potassium permanganate is necessary to replenish the manganese oxides on the medium. The filters should be dosed at a $KMnO_4$ level of 4 to 6 oz (113 to 170 g) per cubic foot of medium. This is best accomplished by mixing the permanganate at a concentration of 4 oz (113 g) to 1 gal (3.8 L) of water. The regenerant should be made up the night before regeneration to ensure that the permanganate is completely dissolved. With this completed, regenerate the filter as outlined below:

1. Shut off inlet-water supply to the filter and depressurize the filter tank by opening a faucet.
2. Drain the water in the filter until it is level with the mineral bed.
3. Add the permanganate solution.
4. Slowly repressurize the filter while running water to the drain until a pink color is noted.
5. Shut off inlet line and drain; allow the filter to stand at least 2 h.
6. Reopen inlet and run water until the pink color disappears.
7. Backwash filter.
8. Purge hose lines to eliminate any excess potassium permanganate.
9. Return the filter to service.

The frequency of regeneration required will depend upon water quality and usage. In ideal situations it may be required as infrequently as once per year or, in worse conditions, as often as once per week. Some manufacturers provide an automatic regeneration. Such systems certainly have advantages but may also present service problems because of the aggressive nature of potassium permanganate. The backwash frequency of iron filters depends upon the type and amount of iron present and the volume of water used. It is suggested that the filter be allowed to run until pressure loss is significant before backwashing. If this takes 6 days, the filter should be backwashed every 4 or 5 days. If this takes 3 days, the filter should be backwashed every 2 days, and so on. Because of this, it is often to the advantage of the user to employ manually backwashed filters and to backwash them only as required. They should be backwashed at the recommended flow rate for 10 to 15 min.

Filters using activated carbon are used to remove undesirable tastes, odors, and organic matter. The primary disadvantage of carbon filters is that, once exhausted, they must be rebedded. The carbon adsorbs foulants, as described in Chap. 9. When the pores of the carbon are laden with contaminants, they can adsorb no more and simply must be replaced in favor of fresh carbon. Unfortunately, there is no simple way of doing this. The filter must be depressurized, the old carbon removed by either vacuuming it out, backwashing it out, or simply dumping it. The new medium is then put into the filter. Unless they are being used for particle removal, carbon filters require backwashing, rather than cleaning to disperse the spent carbon throughout the bed. Backwashing in this type of installation is required less frequently than in cases where turbidity or particulate matter is present.

Clarifying filters are used to remove sand, silt, or other turbidity. These are physical filters and water chemistry is of no concern. The turbidity is trapped at the surface of the bed. As the turbidity builds up, a cake is formed at the surface. Eventually this cake "packs," thereby causing excessive pressure loss. At this point the filter requires backwashing.

Filters used to neutralize acid water contain a form of crushed limestone or magnesia. As described in Chap. 9, the acidic water dissolves the limestone in the process of neutralizing the acid. In practice then, the filter will eventually require replenishing of medium. The pH of the water supply will determine how often this step is necessary. Backwashing of a neutralizer is essential to prevent caking of the bed, and should be done at least once per week.

CARE OF CHEMICAL-FEED PUMPS

The chemical-feed pump is a versatile tool in domestic water treatment. It is also subject to more maintenance than are most other water-treatment devices. This relatively inexpensive item is used to administer some of the most abrasive, corrosive, and active chemicals commercially available. Chlorine, potassium permanganate, sodium hydroxide, sodium carbonate, and acids are only a few of the concoctions pumped into water supplies. To understand the potential problems fully, one must understand the feed pump and how it works. Most feed pumps are positive-displacement diaphragm pumps. They use a diaphragm and cartridge valves as shown in Fig. 11-15.

In addition to the diaphragm and valves illustrated, most pumps also use a strainer or foot valve (see Fig. 11-16) and an injection-point fitting. The strainer serves to keep particulate matter out of the pump. A common complaint is that it gets clogged, which is exactly what it is supposed to do. The injection point is installed into the plumbing line and is used to deliver the chemical to the water supply.

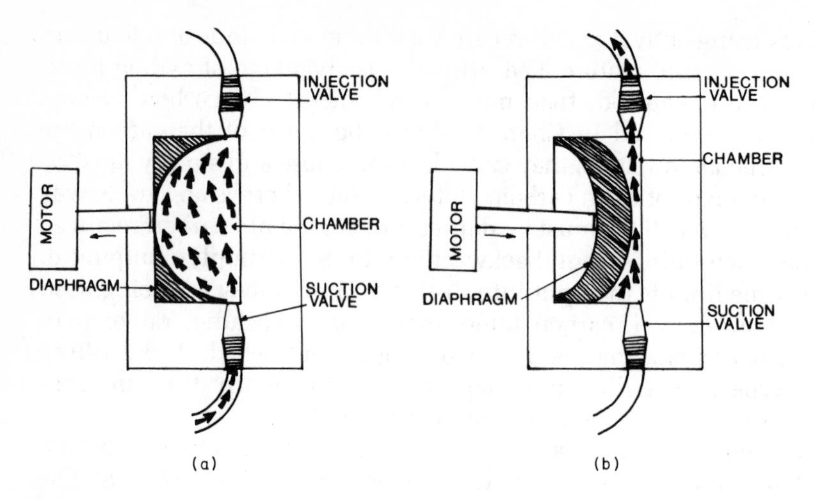

Figure 11-15. Operation of a diaphragm pump. (*a*) Suction cycle: The diaphragm is drawn back by the motor. The suction valve opens, allowing chemical to be drawn from the solution tank. The injection valve is closed, and the chamber is filled with solution. (*b*) Injection cycle: As the diaphragm is driven forward, the suction valve closes and the injection valve opens. The chemical in the chamber is forced out into the injection line.

The potential for problems with chemical-feed pumps stems largely from the eventual attack of the parts by the chemicals being fed. The various valves use metallic springs and synthetic O rings to perform their functions. Such materials are subject to corrosion and degradation by the harsh chemicals mentioned earlier. There is little one can do except minimize the problems by periodically inspecting and replacing the weakened or damaged parts. Table 11-3 can be used as a guide to the use of the chemical feeder:

Table 11-3. Application Guide for Chemical-Feed Pumps

Water condition	*Chemical recommended*	*Additional recommended treatment*
Bacterial contamination	Sodium hypochlorite	Filtration
Iron problem	Sodium hypochlorite Polyphosphates	Filtration
Low pH	Soda ash	
Corrosion problem	Polyphosphates	
Scale control	Polyphosphates	

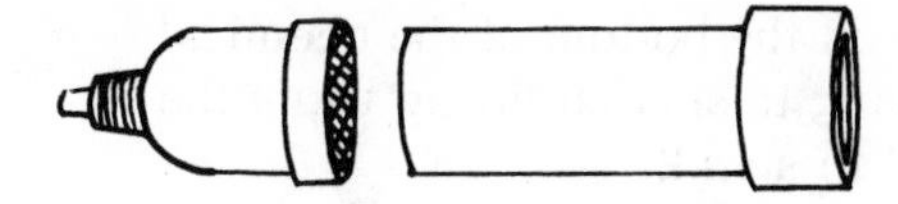

Figure 11-16. Strainer or foot valve.

The chemical-feed pump is somewhat simpler to install than are other types of water-conditioning equipment. It rarely requires any plumbing as such and requires only a small amount of simple electrical wiring. The following steps are offered as guidelines for installing a chemical-feed pump:

1. Place the feeder in an installation location where refitting of the solution tank will be convenient. Also, be sure the location will be warm enough to prevent freezing. The feeder can be placed on a suitable flat surface or can be wall-mounted. *It must be in a fully upright position.*
2. Referring to the installation diagram (Fig. 11-17), attach the strainer (foot valve) to the length of tubing provided. To attach, remove compression nut from strainer and slide over tubing. Then immerse end of tubing in hot water. This will soften the tubing and allow it to accept the tapered form of the fitting. Slide tubing onto the strainer and secure with the compression nut.
3. Run tubing from the foot valve or strainer to the feeder "suction" fitting. Cut tubing at the length needed to hold foot

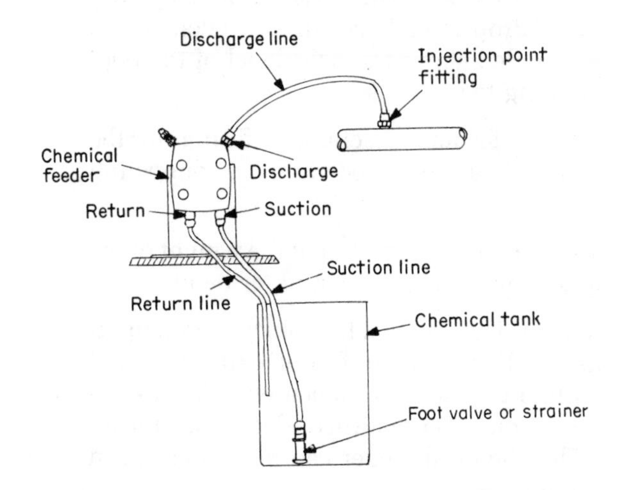

Figure 11-17. Typical chemical-feed-pump installation.

valve or strainer about 2 in off the bottom of the chemical tank as shown. Connect tubing to suction fitting using the same procedure as outlined in step 2.

4. Cut a length of tubing long enough to reach from feeder "return-to-supply" fitting to inside the chemical tank. Return tubing must reach far enough into chemical tank so that it cannot accidentally fall out. Connect tubing to feeder using same procedure as in step 2.

5. Install the Auto-Clean injection-point fitting. Taking into consideration the length of tubing remaining, select the location for the injection fitting. *Do not* install in "dead-ended" pipes or in deeply recessed tees. The tip of the injection fitting must not extend past the centerline of the pipe. *Be sure main system pressure or siphoning will not cause a loss of fluid when installing.* The injection-point fitting has 1/4-in NPT threads. If a like fitting is not available, drill a 7/16-in-diameter hole and carefully thread with a 1/4-in NPT tap. *Do not tap too deeply or the injection fitting will not fit tightly.* Use Permatex or a heavy paint to seal threads on injection point and install fitting. CAUTION: *Do not remove plastic sleeve from tip of the injection fitting. Use an antisiphon valve in the discharge line, if required.* Whenever fluid pressure in the suction line is greater than pressure in the discharge line, the chemical will siphon through the feeder. This can occur if the injection point is on the suction side of a water pump or against a negative head, such as on a submersible or deep-well working-head pump drop line. This can also occur if the chemical tank is installed above the fluid level of the pool or open reservoir being fed.

6. Run tubing from the feeder "discharge" fitting to the injection-point fitting. Use same procedure for connecting as in step 2.

7. Support all tubing in some manner so that the weight of the chemical in tubing cannot cause it to kink or bend.

8. Electrical connections. The voltage, frequency, and amperage requirements are listed on the feeder data plate. All connections must be made in accordance with local electrical wiring codes. Consult a qualified electrician for all internal wiring. The chemical feeder can be connected in the following manner.

Manual operation: The feeder power cord is equipped with the proper plug for the required voltage. Plug the unit into the nearest outlet. Use the switch on the feeder to start and stop pumping action.

Simultaneous operation with another pump or system: Connect the chemical feeder into an outlet *controlled by* the other pump or the system's electrical circuits.

Control by percentage timer: Plug the chemical-feeder electric cord into outlet provided on timer.

Condulet feeders: A separate instruction sheet is included with these models for qualified electricians to follow.

CAUTION: *Be sure chemical feeder is properly grounded to ensure safe operation. Do not turn chemical feeder switch on at this time.*

9. The chemical-feeder relief-release valve is factory-equipped with an 80 lb/in^2 spring. If pressures over 80 lb/in^2 are anticipated, a 115 lb/in^2 spring should be installed. To do this, remove the relief-release valve from the feeder head (Fig. 11-18). Hold the plunger tightly in your fingers and unscrew the knob. Disassemble valve and remove spring. Reassemble using the 115 lb/in^2 spring. Lubricate seals and threads and replace valve into head. Dow Corning 5 compound is recommended for lubricating.

10. Refer to the "Chemical Handling Recommendations," Appendix B. Then fill the chemical storage tank with chemical to be fed.

11. To prime the feeder, turn the knob of the relief-release valve to the open position (Fig. 11-18). Turn the feeder switch on.

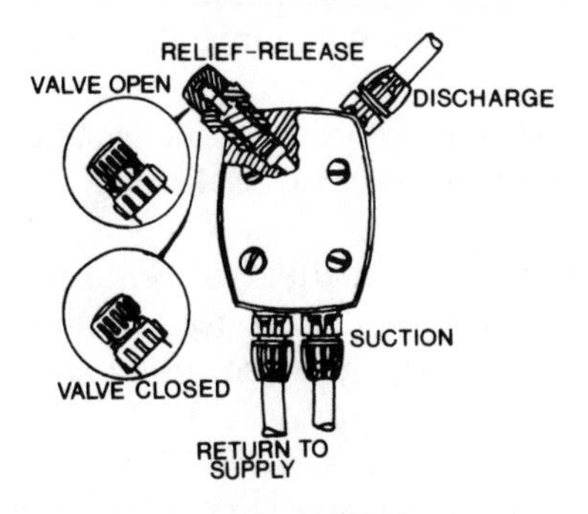

Figure 11-18. Chemical feeder showing details of relief-release valve.

Chemical will be drawn into the head and returned to the chemical tank via the return line. Turn the knob of the relief-release valve back to CLOSE. The feeder is now primed and will feed to the discharge line.

12. Feed-rate adjustment: Refer to the feed-rate charts provided by the manufacturer to determine the feed-rate setting required. To adjust the rate (*be sure feeder switch is on and feeder is operating*), simply turn the adjustment knob to the required setting. CAUTION: *Never attempt to turn this knob when feeder is not operating or damage will result.*
13. When using feeder for chlorination, be sure the detention tank is of adequate size.

Once in operation, certain steps must be taken to ensure continuous service. Probably the most important maintenance step is to be sure to keep the chemical-storage tank full. It should never be allowed to run dry.

Periodically it will be necessary to clean or replace the various valves and fittings on the pump. They can be cleaned by washing non-metallic parts with an acid solution (citric acid, for example). The metallic parts may be washed with water.

CARE OF REVERSE-OSMOSIS EQUIPMENT

Chapter 9 discussed the importance of pretreatment of the water prior to conditioning with a reverse-osmosis (RO) system. The necessity of this pretreatment is so critical that any discussion of RO maintenance must begin with the pretreatment system. The points of concern outlined in Chap. 9 should be incorporated into a maintenance log and all factors should be checked regularly. A complete log is given later in this chapter as Table 11-4.

An RO unit itself is essentially a pump and a module. It consists of prefilters and other incidentals that are actually used for protecting and monitoring the system. See Fig. 11-19.

The prefilter is used to prevent any particulate matter from entering the RO pump and membrane. The first pressure gauge is used to measure the line pressure after the prefilter. This pressure should be at or near the well-system pressure. The RO pressure pump is used to increase the pressure within the membrane and the second pressure gauge is used to monitor the pump pressure. The water is pumped into the membrane and out of the product-water and waste-water lines. The primary maintenance in this area will be replacement of the prefilters and periodic attention required by the pump.

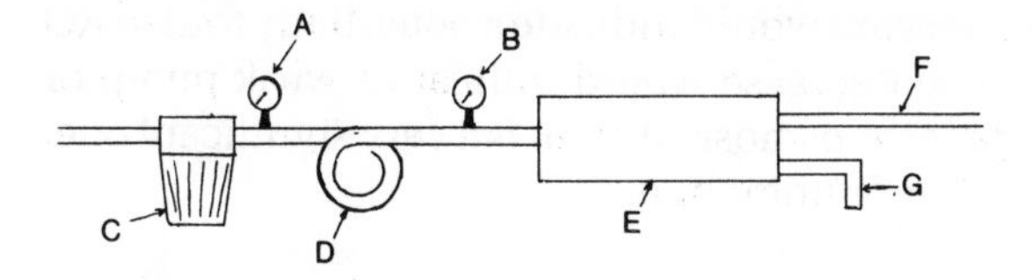

Figure 11-19. A Typical RO unit: A, line-pressure gauge; B, pump pressure gauge; C, prefilter; D, RO pressure pump; E, RO membrane; F, product-water line; G, waste-water line.

A log should be kept on a regular basis (Table 11-4). It can be used to foresee needed maintenance. For example, let's assume the well-system pressure is 40 lb/in^2 (18.14 kg/in^2). Upon installation, the inlet, or filter, pressure is 36 lb/in^2 (16.33 kg/in^2). After 3 weeks, the pressure is 34 lb/in^2 (15.42 kg/in^2). After a month and a half, the pressure suddenly drops to 22 lb/in^2 (9.97 kg/in^2). This information tells us that the prefilter has plugged and requires replacement. Another example would be the pump pressure. Upon installation it is set at 400 lb/in^2 (181.44 kg/in^2). A

Table 11-4. RO System Check List and Maintenance Log

Date	*Inlet hardness*	*Inlet Fe*	*Inlet Cl_2*	*Inlet pressure*	*Pump pressure*	*Prod. flow rate*	*Waste flow rate*	*Inlet TDS*	*Outlet TDS*

sudden increase in pump pressure would indicate a potentially fouled RO membrane whereas a sudden decrease would indicate a weak pump or seals. An investment in a reverse-osmosis system is a very significant one, often equaling the price of an automobile.

SUMMARY

This chapter has attempted to present the basic care of water-conditioning equipment. If it seems vague in places, the author apologizes. To assemble a detailed service manual to cover all makes and models would be a monumental task resulting in an encyclopedia of sorts. Rather than assume that task, the author has attempted to impress upon the reader the advantages and importance of purchasing and/or maintaining domestic water-conditioning equipment.

REFERENCES

American Association for Vocational Instructional Materials, 1973: *Planning for an Individual Water System*, Athens, Georgia.

American Water Works Association, 1971: *Basic Water Treatment Operator's Manual*, New York.

American Water Works Association, 1971: *Water Quality Treatment*, 3d ed., McGraw-Hill, New York.

U.S. Environmental Protection Agency, Office of Water Supply, 1977: *State of the Art of Small Water Treatment Systems*, Washington, D.C.

12

Water-Treatment Waste Disposal

Waste waters generated by households are either transported by sewers to central facilities for treatment and disposal or treated and disposed of on-site by some type of septic system. In 1970, nearly 29 percent of the nation's housing units (about 19.5 million homes) used on-site disposal systems; each year about 500,000 of the new homes built are equipped with on-site systems.

In recent years, many Americans have been moving to suburban residential areas where domestic waste-water disposal is handled largely by on-site systems. The most common type of system (comprising 85 percent of all on-site units) is the septic tank.

A recent report by the General Accounting Office (1978) indicates that septic systems are environmentally sound, technologically feasible, and cost-effective. The report adds that septic systems, if properly handled, can be a permanent method of waste treatment.

Studies have attempted to determine the exact nature of the waste effluent from recharging water softeners as well as its effect on private sewage-disposal systems. Three major areas have been examined, all of which deal with the effect of effluents generated during the recharge cycle of household water softeners.

Study of the effect of dissolved salts in softener-recharge effluent on biological action in septic-tank systems demonstrated that recharge effluent from water softeners had no deleterious effect on the biological action in a septic tank. It was found that recharge waste effluents may actually stimulate biological action (Weibel et al., 1955).

The hydraulic effect of the volume of water-softener waste water on biological action has been researched. Findings indicate that the volume of recharge effluent from a water softener is less than that from present-day automatic clothes washers (Weickart, 1976). The amount of waste effluent developed by a typical household water softener during recharge is about 50 gal (200 L); the effluent contains calcium, magnesium, and sodium chlorides. The frequency of recharge is dependent on water hardness, usage, and recharge salt dosage.

SEPTIC-TANK SYSTEMS

Another area of concern is the effect of softener-recharge effluents on soil percolation in septic-system drain fields. Since much literature on irrigation contains references to the adverse effects of high-sodium water on soil structure and permeability (particularly in clay-type soils), this area is of particular significance. Two studies were conducted to investigate this subject. One dealt solely with anaerobic septic-tank systems, and the other with aerobic treatment systems (National Sanitation Foundation, 1978). Researchers concluded that there is an important difference between water-softener effluents and sodium effluents (Tyler et al., 1978).

Water-softener effluents contain significant amounts of calcium and magnesium, so they are not really just sodium effluents. Calcium and magnesium counteract the effect of sodium and help maintain and sustain soil permeability, even in susceptible clay-type soils. Thus, it appears that the effluent brine from water-softener recharge will not affect biological digestion, hydraulic load, or leach-field permeability in a septic-tank system. However, if the leach field is composed of swelling clays, permeability will be reduced regardless of the presence of water-softener effluent.

Salts in the waste effluent from recharge of water softeners created no percolation problems in septic-tank seepage fields, according to another study. In fact, water-softener recharge effluents increased soil percolation. In other words, lower permeability may result if regeneration brines from water softeners are not allowed to enter the septic-tank seepage field. If the regeneration wastes are discharged to a dry well, a roadside ditch, or a point other than the septic system, the beneficial effects of calcium and magnesium will be lost.

WASTE-DISPOSAL METHODS

Aerobic Systems

Certain substances, such as toilet-bowl cleaners, may prove toxic to the bacteria in aerobic systems when they are discharged into the tank. This results in slower waste treatment. However, early studies by Ludzack and Noran (1965) and recent findings of the National Sanitation Foundation (1978) indicate that regeneration brines do not retard biological action in aeration systems.

Wells

Shallow wells can become contaminated by water-softener regeneration brines leaking from leach fields. If the cone of depression in the water table surrounding a pumping well intercepts the recharge-effluent mound beneath a septic-tank leach field, regeneration brine may contaminate the well (Fig. 12-1).

Local hydrology is important. A well contaminated by water-softener brine should be investigated immediately. Though chloride alone does not pose a health threat, its appearance in a water supply suggests the presence of other septic-tank substances, such as nitrate and pathogenic organisms which are potentially harmful. Water should be tested for nitrate, sulfate, detergents, and coliform bacteria.

Dry Wells

Dry wells are shallow excavations filled with rock. Fluids discharged into them may seep into the ground and eventually reach nearby shallow wells, particularly in sandy strata. Deep wells are usually not contaminated by dry-well seepage. However, dry-well disposal should always be used with caution.

Land Surface and Ditches

While such disposal is rare, insufficiently diluted water-treatment wastes can kill vegetation and pollute surface-water supplies if they are

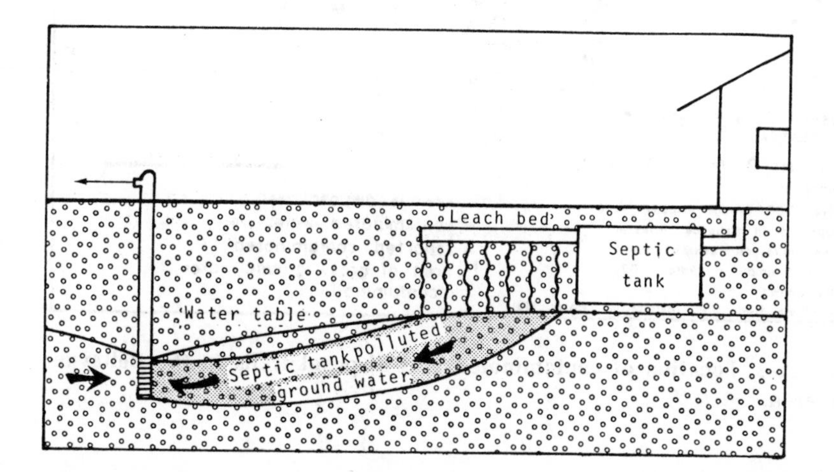

Figure 12-1. Percolation through zone of aeration. Most of the natural removal or degradation processes function under these conditions.

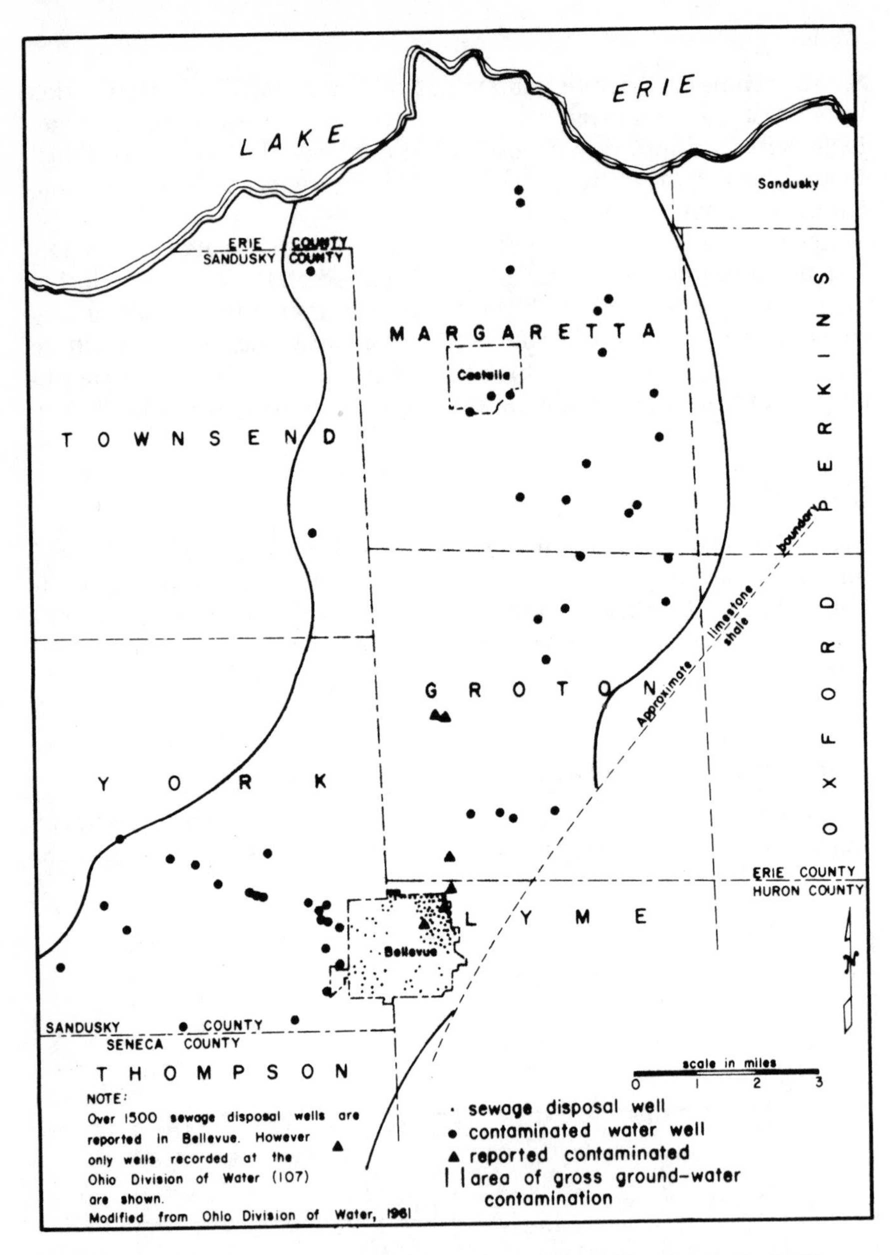

Figure 12-2. Location of contaminated and disposal wells in the Bellevue, Ohio, area.

allowed to flow directly onto the land surface. These wastes may also contaminate shallow ground water reservoirs and wells.

Such wastes can often be discharged safely, however, in arid regions where the soil permeability is low and aquifers are well protected by overlying material [at least 30 ft (10 m) below land surface].

Disposal Wells and Sinkholes

Sinkholes, which appear in limestone, may tap a vast network of underground caves and tunnels. Septic-system wastes should never be discharged into shallow disposal wells (particularly abandoned water wells), since they may flow for miles without filtration or degradation.

Bellevue, Ohio, experienced the deleterious effects of sewage effluent wastes discharged into disposal wells and sinkholes. A highly permeable limestone aquifer was contaminated by household, municipal, and industrial wastes that were allowed to flow into scores of sinkholes and drilled wells. Septic-tank overflow was also drained into the wells. Sewage effluent contaminated the ground water over an area that exceeded 75 mi^2 (Fig. 12-2).

Streams

Raw sewage and septic-tank effluent discharged into streams can be a direct source of surface water contamination. These wastes may harm stream-water quality, particularly during late-summer, low-flow conditions. Federal and state laws prohibit this once-common disposal practice.

REFERENCES AND BIBLIOGRAPHY

Brady, N. C., 1974: *The Nature and Properties of Soils*, 8th ed., Macmillan, New York, pp. 59–60.

General Accounting Office, 1978: *Community-Managed Septic Systems—A Viable Alternative to Sewage Treatment Plants*, CED-78-165.

Ludzack, E. J. and D. K. Noran, 1965: "Tolerance of High Salinities by Conventional Wastewater Treatment Processes," *J. Water Pollut. Contr. Fed.*, vol. 37, pp. 1404–1406.

National Sanitation Foundation, 1978: *The Effects of Home Water Softener Waste Regeneration Brines on Individual Aerobic Wastewater Treatment Plants*, Water Quality Research Foundation, Lombard, Illinois.

Otis, R. J., N. J. Hutzler, and W. C. Boyle, 1975: *On-site Household Wastewater Treatment Alternatives—Laboratory and Field Studies in Water Pollution*

Control in Low Density Areas, W. J. Jewell and R. Swan (eds.), University of Vermont, Burlington, pp. 241–265.

Rudolfs, Willem, 1928: "Effect of Salt on Sludge Digestion," U. S. Public Health Service Reprint 1221.

Tyler, E. J., R. B. Corey, and M. U. Olotu, 1978: *Potential Effects of Water Softener Used on Septic Tank Soil Absorption On-site Waste Water Systems*, Small Scale Waste Water Management Project, University of Wisconsin, Madison.

Weibel, S. R., T. W. Bedixen, and J. B. Coulter, 1955: *Studies on Household Disposal Systems, Part III*, U.S. Public Health Service, U.S.D.A., pp. 47–48.

Weickart, R. F., 1976: *Backwash Water and Regeneration Wastes from Household Water Conditioning Equipment on Private Sewage Disposal Systems*, Water Quality Association, Lombard, Illinois, p. 8.

APPENDIX A
State Agencies That Provide Water-Sampling Services

State	*Private labs?*	*Certified*	*Where should homeowner go first?*	*Name, address, phone number for more information*
Alaska	Y	. . .	Regional laboratory	Southeast Regional Laboratory, Pouch J, Juneau, AK 99811 (907)586-3586
Alabama	Y	. . .	County health	Division of Public Water Supplies, 560 S. McDonough St., Montgomery, AL 36130 (205)832-3170
Arizona	Y	Y	Private laboratory	Department of Health Services, Bureau of Water Quality Control, 1740 W. Adams St., Phoenix, AZ 85007 (602)271-5453
Arkansas	Y	N	County health	State Department of Health, 4015 W. Markham, Little Rock, AR 72201 (501)661-2060
California	Y	Y	District office	District Office of Sanitary Engineering or State Department of Health, Sanitation Engineering Section, 2151 Berkeley Way, Berkeley CA 95814 (916)445-1576
Colorado	Y	Y	County health	Colorado Department of Health, 4210 East 11th, Denver, CO 80220 Attn.: Water Quality Laboratory (303)388-6111
Connecticut	Y	Y	Private laboratory	Connecticut Department of Public Health, Hartford, CT 06115 (203)566-2211

State	Private labs?	Certified	Where should home-owner go first?	Name, address, phone number for more information
Delaware	Y	Y	County health	Local county health laboratory
Florida	Y	N	County health	Local county health department, or Health and Rehabilitation Services, Division of Health, Tallahassee, FL 32301 (904)488-3153
Georgia	Y	N	State	Department of Human Resources, Environmental Health Section, 41 Trinity Ave., Atlanta, GA 30334 (404)656-4807
Hawaii	One	N	State	Sanitation Branch, Department of Health, 1250 Punchbowl St., Honolulu, HI 96813 (808)548-6478
Idaho	Y	. . .	Central health district	Bureau of Water Quality, Statehouse, Boise, ID 83720 (208)384-2287
Illinois	Y	Y	State or county health boards	Illinois Department of Public Health Laboratories, 134 N. 9th St., Springfield, IL 62701 (217)782-6562
Indiana	Y	Y	State	Indiana State Board of Health, Water and Sewage Laboratory, 1330 W. Michigan St., Indianapolis, IN 46206 (317)633-0232
Iowa	Y	Y	State	State Hygienic Laboratory, Medical Laboratory Building, Iowa City, IA 52242 (515)281-5605
Kansas	Y	N	County or regional health	Office of Laboratories and Research, Kansas Department of Health and Environment, Forbes Building 740, Topeka, KS 66620 (913)862-9360
Kentucky	Y	Y	County health	Department of Natural Resources, Water Laboratory, Frankfort, KY 40601 (502)564-4446 (bacteriological), 564-3772 (chemical)
Louisiana	Y	N	Parish health	Louisiana Office of Health Services and Environmental Quality, Bureau of Environmental Services, Jane F. Kerber, Director, P.O. Box 60630, New Orleans, La 70160 (504)568-5101
Maine	Y	Y	State health	Public Health Laboratories, Statehouse, Augusta, ME 04333 (207)289-1110

State	Private labs?	Certified	Where should homeowner go first?	Name, address, phone number for more information
Maryland	Y	N	County health	County health departments
Massachusetts	Y	Y	Local health	Division of Water Supply, 600 Washington St., Room 320, Boston, MA 02111 (617)727-2692
Michigan	Y	Y	County health	Sanitary Bacteriological and Chemical Laboratories, Bureau of Disease Control and Laboratory Services, Michigan Department of Public Health, Box 30035, Lansing, MI 48909 (517)373-1428
Minnesota	Y	Y	State health	Minnesota Department of Health, Analytical Laboratory, Room 405, 717 Delaware St. SE, Minneapolis, MN 55440 (612)296-5335
Mississippi	Y	Y	County health	County health departments
Missouri	Y	N	District health	Bureau of Community Sanitation, 1407 SW Blvd., Jefferson City, MO 65101 (314)751-2151
Montana	Y	Y	State	Water Quality Bureau, Cogswell Building, Capitol Station, Helena, MT 59601 (406)449-2407
Nebraska	Y	N	County agent	County Cooperative Extension Service
Nevada	Y	Y	State	Consumer Health Protection Services, Room 103, Kinkead Building, Capitol Complex, Carson City, NV 89710 (702)784-4750
New Hampshire	Y	N	State	Water Supply and Pollution Control Laboratories, Hazen Dr., Concord, NH 03301 (603)271-3445
New Jersey	Y	Y	Private laboratories	Bureau of Potable Water, Division of Water Resources, 1474 Prospect St., Trenton, NJ 08625 (609)292-2121
New Mexico	Y	Y	EIA (regional)	Environmental Improvement Agency, Central Office, 725 St. Michael's Dr., Crown Building, Santa Fe, NM 87503 (505)827-4011
New York	Y	Y	County health	Division of Laboratories and Research, Empire State Plaza, Albany, NY 12237 (518)474-3968

State	*Private labs?*	*Certi-fied*	*Where should home-owner go first?*	*Name, address, phone number for more information*
North Carolina	Y	Y	County health	County health departments
North Dakota	Y	N	District health	Public Health Laboratories, Box 1678, Bismarck, ND 58501 (701)224-2304
Ohio	Y	Y	Municipal health or county health	Ohio EPA, Box 1049, 361 E. Broad St., Columbus, OH 43216 (614)466-8565
Oklahoma	Y	Y	County health	County health departments or (405)271-5220 (Bacteriological and (405)271-5240 (Chemical)
Oregon	Y	Y	County environ-mental health	Water Resources Department, 555 13th Street, Mill-Creek Office Park, Salem, OR 97310 (503)378-8455
Pennsylvania	Y	Y	Com-munity Environ-mental Control Depart-ment of Environ-mental Research	Bureau of Community Environmental Control, Harrisburg, PA 17120 (717)787-9037
Rhode Island	Y	Y	State health	State of Rhode Island Health, Water and Wastewater Laboratory, State Offices, Providence, RI 02903 (401)277-2968
South Carolina	Y	Y	County/ district health	EQC Water Laboratory, South Carolina Department of Health and Environmental Control, 2600 Bull St., Columbia, SC 29201 (803)758-5496
South Dakota	N	N	County health	South Dakota Department of Environ-mental Protection, Office of Water Hygiene, Foss Building, Pierre, SD 57501 (605)244-3754
Tennessee	Y	Y	County health	Division of Local Health Services, Department of Public Health, Attn.: James Ault, Nashville, TN 37219 (615)741-2275

State	*Private labs?*	*Certi-fied*	*Where should home-owner go first?*	*Name, address, phone number for more information*
Texas	Y	N	Regional health	Division of Water Hygiene, Texas Department of Health, 1100 W. 49th St., Austin, TX 78759 (512)475-2323
Utah	Y	Y	Local health district	Local health district offices
Vermont	. . .	. . .	Local health depart-ment	Vermont State Department of Health, 60 Main Street, Burlington, VT 05401 (802)862-5701
Virginia	Y	N	County health	County Health Departments or Division Consolidated Laboratories, Bureau of Microbiology, Richmond, VA 23230 (804)786-3756
Washington	Y	N	Private laboratories or county health	Department of Social and Health Services, Water Supply and Waste Section, P.O. Box 1788, Olympia, UT 98504 (206)464-7670
West Virginia	Y	. . .	County health	Bacteriological: State Hygienic Laboratory, 167 11th Ave., S. Charleston, WV 25363 (304)348-8143 Chemical: Environ-mental Health Services Laboratory, Attn.: James Rosencrance, 151 11th Ave., S. Charleston, WV 25363 (304)348-0197
Wisconsin	Y	Y	State health	State Laboratory of Hygiene, 460 Henry Mall, Madison, WI 53706 (608)266-2211
Wyoming	Y	Y	State health	Bacteriological: Public Health Laboratory Services, Wyoming Department of Health and Social Services, Hathaway Building, Cheyenne, WY 82002 (307)777-7781 Chemical: Wyoming Department of Agriculture, State Chemistry Laboratory, Box 3228, Laramie, WY 82071 (307)766-3381

APPENDIX B
Chemical Handling Recommendations

Some of the chemicals used in conjunction with water-treatment equipment, such as the chemical-feed pump, present potential hazards. Some are explosive and others can prove fatal if inhaled or ingested. The Chemical Handling Recommendations* below serve as a guide to the safe use and storage of selected chemicals.

CALCIUM HYPOCHLORITE

(Bleaching Powder) $Ca(ClO)_2$

Description: White powder, granules, or pellets with strong chlorine-like odor.

Fire and explosion hazards: Powerful oxidizing material. With acids evolves chlorine at ordinary temperatures (*see* Chlorine). Not combustible, but evolves oxygen at higher temperatures. Readily ignites combustible or organic materials when in contact. May undergo accelerated decomposition with evolution of heat.

Life hazard: Irritating to skin, eyes, and respiratory tract.

Personal protection: Wear self-contained breathing apparatus; wear goggles if eye protection not provided.

Fire-fighting phases: Use water, preferably in form of spray.

Usual shipping containers: Airtight cans, wooden barrels, steel drums, fiber drums.

Storage: Protect against physical damage. Store in cool, dry, well-ventilated place away from combustible materials. Drums may rupture from exposure to heat, particularly if chlorine content is high. Avoid storage for prolonged periods, particularly at summer temperatures.

Remarks: See Code for the Storage of Liquid and Solid Oxidizing materials (NFPA No. 43A).

* *Source: Fire Protection Guide on Hazardous Materials,* 5th ed., 1973, National Fire Protection Association, Boston, Massachusetts.

HYDROGEN CHLORIDE

(Hydrochloric Acid) HCl

Description: Hydrogen chloride is a colorless gas. Hydrochloric acid, a water solution of hydrogen chloride, is a clear, colorless or slightly yellow, fuming liquid with an irritating pungent odor.

Fire and explosion hazards: Not combustible but contact with common metals produces hydrogen which may form explosive mixtures with air. Soluble in water.

Life hazard: Toxic. Eye, skin, and respiratory irritant. Inhalation of concentrations of about 1500 ppm in air is fatal in a few minutes.

Personal protection: Wear full protective clothing.

Fire-fighting phases: Use water, neutralize with chemically basic substances such as soda ash or slaked lime.

Usual shipping containers: Aqueous solutions in glass bottles, carboys, rubber-lined tank cars. Anhydrous hydrogen chloride in steel cylinders and tank barges.

Storage: Protect against physical damage. Store in cool, well-ventilated place, separated from all oxidizing materials.

Remarks: See Chemical Safety Data Sheet SD-39 (Manufacturing Chemists' Association, Inc.)

POTASSIUM PERMANGANATE

(Iron Filter Regenerant) $KMnO_4$

Description: Dark-purple crystals with blue metallic sheen.

Fire and explosion hazards: Powerful oxidizing material. Explosive in contact with sulfuric acid or hydrogen peroxide. Reacts violently with finely divided, easily oxidizable substances. Increases flammability of combustible materials.

Life hazard: Skin and eye irritant.

Fire-fighting phases: Use water.

Usual shipping containers: Bottles and cans

Storage: Protect against physical damage. Separate from sulfuric acid, hydrogen peroxide, and all combustible, organic, or readily oxidizable materials.

SODIUM CHLORITE

(Bleach) $NaClO_2$

Description: White crystals or crystalline powder.

Fire and explosion hazards: Very powerful oxidizing material. Decomposes at 347°F (175°C) with evolution of heat. Forms explosive mixtures with combustible organic or other readily oxidizable materials. In contact with strong acids, releases explosive chlorine dioxide gas. Containers may explode when involved in fire.

Life hazard: Toxic. Avoid excessive skin contact with solid or solutions. Avoid repeated or prolonged inhalation. Dangerous in contact with acids, releasing extremely poisonous chlorine dioxide gas.

Fire-fighting phases: Use water.

Usual shipping containers: Wooden boxes with inside glass or metal containers, or metal containers up to 100-lb (45 kg) capacity.

Storage: Protect against physical damage. Store in cool, dry place, preferably in detached fire-resistive building. Separate from combustible, organic or other readily oxidizable materials, acids, sulfur and flammable vapors. Immediately remove and carefully dispose of any spilled chlorite.

Remarks: See Code for the Storage of Liquid and Solid Oxidizing Materials (NFPA No. 43A).

SODIUM HYDROSULFITE

(Resin Cleaner) $Na_2S_2O_4$

Description: White to grayish crystalline powder; may have faint sulfurous odor.

Fire and explosion hazards: Combustible solid but not explosive. Burns slowly, about like sulfur. Heats spontaneously in contact with moisture and air; may ignite nearby combustible materials.

Life hazard: An irritant. Burning produces extremely toxic sulfur dioxide gas.

Personal protection: In fire conditions wear self-contained breathing apparatus.

Fire-fighting phases: Avoid use of water unless flooding amounts are available for application and flushing; carbon dioxide, dry chemical, and dry sand are best.

Usual shipping containers: Wooden barrels, kegs, or boxes with inside glass bottles or metal containers, not exceeding 5 lb (2.25 kg) each; specially constructed fiber or metal drums.

Storage: Protect against physical damage and moisture. Separate from combustible organic or other readily oxidizable substances.

SODIUM HYDROXIDE

(Lye) NaOH

Description: White pellets, flakes, sticks, or a solid cast mass completely filling a drum; also may be a concentrated water solution.

Fire and explosion hazards: Not combustible, but solid form in contact with moisture or water may generate sufficient heat to ignite combustible materials. Contact with some metals can generate hydrogen gas.

Life hazard: Toxic. A severe eye hazard; solid or concentrated solution destroys tissue on contact.

Personal protection: Wear full protective clothing.

Fire-fighting phases: Flood with water, using care not to splatter or splash this material.

Usual shipping containers: Bottles, cans, drums, tank cars, tank barges.

Storage: Protect against physical damage. Store in dry place; protect against moisture and water. Separate from acids, metals, explosives, organic peroxides, and easily ignitable materials.

Remarks: See Chemical Safety Data Sheet SD-9 (Manufacturing Chemists' Association, Inc.).

APPENDIX C
Glossary

Absorption The process by which one substance is taken into the body of another substance.

Acid A substance which releases hydrogen ions when dissolved in water. A strong acid will release a large proportion of hydrogen ions whereas a weak acid will release a small proportion of hydrogen ions.

Acidity The ability of a water solution to neutralize an alkali or base.

Acid-Mine Drainage A nondescript term indicating a stream carrying leachate from a mine, usually a strip mine, characterized by low pH, high iron content, and high dissolved solids.

Actinomycetes A soil bacterium, notable for musty state and odor.

Activated Carbon A material that has a very porous structure and is an adsorbent for organic matter and certain dissolved gases.

Adsorption The process by which a gas, vapor, dissolved material, or very tiny particle adheres to the surface of a solid.

Aeration The process by which air becomes dissolved in water, usually by spraying water into the air or by bubbling air through water.

Aerobic An action or process conducted in the presence of oxygen.

Algae Bloom *See* water bloom.

Alkalinity The ability of a water solution to neutralize an acid.

Alum A common name for aluminum sulfate, used as a coagulant.

Ammonia Fertilizer A material with a high concentration of nitrogen compounds, put on soil to stimulate plant growth.

Anaerobic An action or process conducted in the absence of oxygen.

Angstrom (Å) A unit of length equal to one-ten-thousandth of a micrometer, or one-tenth of a millimicron. 1 Å = 1×10^{-10} m; 1 Å = 1×10^{-8} cm.

Anion An ion with a negative electric charge.

Aquiclude A body of relatively impermeable rock that is capable of absorbing water slowly but functions as an upper or lower boundary of an aquifer and does not transmit ground water rapidly enough to supply a well or spring.

Aquifer A body of rock, or a zone within a body of rock, that contains sufficient saturated permeable material to yield economically significant quantities of ground water to wells and springs.

Arsenosis An affliction caused by the accumulation of arsenic in the body.

Artesian Aquifer An aquifer bounded above and below by impermeable beds or beds of distinctly lower permeability than that of the aquifer itself, causing the water to be under a hydrostatic pressure greater than the lithostatic pressure.

Artificial Recharge The practice of increasing, by artificial means, the amount of water that enters a ground water reservoir.

Atom The smallest particle of an element which retains characteristics of the element.

Atomic Mass Unit The standard mass unit defined by international agreement as one-twelfth the mass of the carbon isotope (C 12) with six neutrons.

Atomic Number The number of protons within the nucleus of an atom. All atoms of a given element have the same atomic number.

Atomic Weight A number representing the weight of one atom of an element as compared with an arbitrary number representing the weight of one atom of another element taken as the standard (*see* atomic mass unit).

Backwash The process in which beds of filter or ion-exchange media are subjected to flow opposite to the service-flow direction to loosen the bed and to flush suspended matter, collected during the service run, to waste.

Bacteria Typically, unicellular microorganisms which have no chlorophyll and multiply by simple division. Some bacteria cause disease, but others are necessary for fermentation, nitrogen fixation, etc.

Baffle A deflector at the inlet (top) of a typical water softener which disperses the water over the top of the resin bed.

Bailer A long, hollow, cylindrical steel container or pipe with a valve at the bottom, attached to a wire line and used in cable-tool drilling for recovering and removing water, cuttings, and mud from the bottom of a borehole or well.

Base A substance which releases hydroxyl ions when dissolved in water. A strong base will release a large proportion of hydroxyl ions, whereas a weak base will release a small proportion of hydroxyl ions.

Biocide A substance that is destructive to many different organisms.

Biological–Biochemical Degradation The breakdown of a material (sewage, refuse, etc.) by the action of organisms, bacteria, and weathering.

Biological Oxygen Demand (BOD) A measure of the amount of oxygen consumed in the oxidation of organic matter by biological action, most often in reference to the strength of waste water or sewage.

BOD *See* biological oxygen demand.

Bonding The means by which atoms or groups of atoms are combined in molecules.

Bore Well A shallow (3 to 30 m, or 10 to 100 ft), large-diameter (20- to 90-cm, or 8- to 36-in) water well constructed by hand-operated or power-driven augers.

Brine A concentrated saline solution.

Cable-Tool Drilling A method of drilling based on a percussion principle in which the rock material at the bottom of the hole is pulverized or broken up by means of a solid-steel cylindrical bit attached to, and working vertically at, the end of a steel cable activated by a walking beam, the bit chipping the rock by a series of repeated blows.

Carbonate A mineral compound characterized by a fundamental anionic structure of CO_3^{2-}. When a carbonate mineral such as calcite is dissolved in water, the amount remaining as CO_3^{2-} is measured as carbonate.

Carbonate Hardness Hardness due to the presence of calcium and magnesium bicarbonates and carbonates in water. When the hardness is numerically greater than the sum of the carbonate alkalinity and the bicarbonate alkalinity, the amount of hardness that is equivalent to the total alkalinity is carbonate hardness.

Casing A heavy metal or strong plastic pipe or tubing of varying diameter and weight lowered into a borehole during or after drilling in order to support the sides of the hole and thus prevent the walls from caving in, the loss of drilling mud into the porous ground, and water, gas, or other fluid from entering the hole.

Cathartic A medicine that stimulates evacuation of the bowels, a laxative.

Cation An ion with a positive electric charge.

Cesspool A tank or deep hole in the ground for receiving drainage or sewage from a residence. More recently used to describe a lake which has been befouled by raw sewage or sewage-treatment effluent.

Chemical-Feed Pump A mechanical device designed to introduce chemicals into a water system at a rate proportional to the water flow. Also called a *chemical feeder.*

Chlorination The use of chlorine gas or solutions of its compounds to disinfect water or as an oxidizing agent.

Chlorine Demand A measure of the amount of chlorine which can be consumed by organic matter and other oxidizable substances in water without a chlorine residual.

Cistern An artificial reservoir or tank for storing water.

Clarification The process of making water clear and free of suspended impurities.

Clay A rock or mineral fragment or a detrital particle of any composition smaller than a very fine silt grain, having a diameter less than $^1/_{256}$ mm.

Coagulation The process in which very small, finely divided solid particles, often colloidal in nature, are agglomerated into larger particles.

Compound A chemical combination of two or more elements in definite ratios by weight in which the set of characteristics of each element is lost.

Concentrated Describes a solution which contains a relatively large quantity of solute.

Conductance A measure of the ability of a solution to carry an electric charge; the reciprocal of the electric resistance.

Conductivity The quality or power of a medium to transmit electrical charges; in water, the conductivity is related to the concentration of ions capable of holding electrical charges.

Cone of Depression The area of influence of a well. The extent of the lowering of the potentiometric surface around a well.

Confined Aquifer *See* artesian aquifer.

Confining Bed A body of impermeable or distinctly less permeable material stratigraphically adjacent to one or more aquifers. Also called an *aquiclude.*

Consumptive Use The difference between the total quantity of water withdrawn from the source for use and the quantity of water, in liquid (and, rarely, solid) form, returned to the source. It includes mainly water transpired by plants and evaporated from the soil.

Containment Structure *See* holding structure.

Covalent Bond The bond formed by the sharing of a pair of electrons (i.e., H_2).

Crenothrix polyspora A species of filamentous bacteria which utilizes iron in metabolism. One of many iron bacteria.

Crib A framework to support the walls of a pit, such as a dug well, or to protect them from caving.

Dechlorination The removal of excess chlorine residual.

Desulphovibrio desulfuricans A common sulfate-reducing bacterium causing accelerated corrosion of pipes.

Detergent Any material with cleansing powers, including soaps, the newer synthetic detergents, many alkaline materials, and solvents. It strictly refers to synthetic detergents such as ABS or LAS (alkyl benzene sulfonate, linear alkyl sulfonate).

Deuterium An isotope of hydrogen containing one neutron and having a mass of 2.0141 u. Also called *heavy hydrogen*.

Diatom A microscopic, single-celled organism growing in marine or fresh water. Diatoms secrete a skeleton of siliceous material in a great variety of forms that may accumulate in sediments in enormous numbers.

Diatomaceous earth A white, yellow, or light-gray siliceous earth composed predominantly of the remains of diatoms that typically accumulated in lakes or swamps. It is used as an absorbent and filtering agent.

Dilute Describes a solution which contains a relatively small quantity of solute.

Disposal Well A well used to dispose of waste by injection into a deep aquifer.

Dissolved Solids The weight of matter in true solution in a stated volume of water, including both inorganic and organic matter.

Distributor A device or system designed to produce even flow through all sections of an ion exchanger or filter bed and to retain the filter medium in the tank or vessel.

Diuretic Something tending to cause an increase in the flow of urine.

Domestic A term applying to a household or private residence.

Drilled Well A well constructed by either cable-tool or rotary methods usually to depths exceeding 50 ft with capacities to provide for industry, irrigation, or municipalities.

Driven Well A shallow, usually small-diameter (3- to 10-cm) well constructed by driving a series of connected lengths of pipe into unconsolidated material to a water-bearing stratum, without the aid of any drilling, boring, or jetting device.

Dry Well A well that is dry (usually one that at one time was producing) as a result of the decline of the water table or potentiometric surface.

Dug Well A shallow, large-diameter well constructed by excavating with hand tools or power machinery instead of drilling or driving, typically for individual domestic water supplies and yielding considerably less than 100 gal/min (380 L/min).

Dump See landfill.

Effluent A liquid discharged as waste, such as contaminated water from a factory, outflow from a sewage-treatment facility, or storm-sewer discharge.

Also, the finished treated-product water coming from a water-treatment device.

Electric Conductance A measure of the ability to permit the flow of electric charge, related to the concentration of ions in water.

Electron A negatively charged particle that revolves around the nucleus of an atom and has a mass equal to approximately $^1/_{1836}$ of the mass of a proton.

Electroneutrality The occurrence of zero net charge in any ionic solution.

Element Any substance that cannot be separated into different substances by ordinary chemical methods. Each element has its own set of characteristics, which differs from that of all the other elements.

Encrustation A covering of crust or crustlike layer on the surface of an object.

Equilibrium The state in which the action of multiple forces produces a steady balance, resulting in no change over time.

Equivalent weight The weight, in grams, of an element, compound, or ion which would react with or replace one gram of hydrogen; the molecular weight in grams divided by the valence.

Evaporation The process by which a substance changes from the liquid to the vapor state.

Evaporite A mineral precipitated as a result of evaporation.

Excavation A pit or mine made by human beings.

Exploratory Well A well drilled in unproven territory either in search of a new and as yet undiscovered field of oil or gas or water or with the expectation of extending the known limits of a field or aquifer already partly developed.

Fabric Softener A common additive in laundering which softens water to increase the efficiency of the soap or synthetic detergent.

Fecal Coliform Matter containing or derived from animal or human waste containing one or more of the coliform groups of bacteria.

Fecal Streptococcus Matter containing or derived from animal or human waste containing one or more of the streptococcus groups of bacteria.

Field Capacity The amount of water held in against the pull of gravity by capillary or molecular attraction. Expressed in percent. Also called *specific retention.*

Flavor The combined sensation of taste, odor, temperature, and feel.

Float-Type Filter A water intake used in cisterns which is held some distance below the surface by a float and has a filter to prevent suspended matter from entering the system.

Flocculation The agglomeration of finely divided suspended solids into larger, usually gelatinous, particles.

Flow Diverter A device which prevents initial precipitation from flowing directly into a cistern and permits the roof or catchment area to be washed of dust and debris. Also called a *roofwasher.*

Fluorescein An orange-red compound, $C_{20}H_{12}O_5$, that exhibits intense fluorescence in alkaline solution and is used to dye water in order to trace its movement.

Fluoride A general reference to compounds containing fluorine used to supplement water supplies for reduction of dental caries.

Fluorosis An affliction causing discolored teeth and in severe cases pitting of the enamel; caused by excessive concentrations of fluoride in drinking water.

Foot Valve *See* strainer.

FWPCA The Federal Water Pollution Control Act.

Gallionella ferruginea A species of stalked, ribbonlike bacteria which utilize iron in their metabolism. One of many types of iron bacteria.

Grain A unit of weight equal to $^1/_{7000}$ pound, or 0.0648 gram.

Ground Water Water in the saturated zone found under hydrostatic pressure.

Grouting A cementitious fluid poured or injected into a borehole during well drilling to seal crevices and prevent contamination or the loss of drilling mud, to provide a protective wall around the metal casing, or to improve the strength and elastic properties of the rock.

Halogen Any of the elements of group VIIA on the periodic table from the Greek term meaning "salt formers." Halogens are the most reactive nonmetallic elements. They are capable of reacting with practically all metals and most nonmetals, including themselves. The halogens include fluorine, chlorine, bromine, iodine, and astatine.

Hardness A measure of soap-neutralizing ions present in water; predominantly magnesium and calcium, but other alkali metal ions contribute to the effect.

Holding Structure Any excavation, pond, or closed embankment used to contain water or another solution until needed.

Humidifier A mechanical device for increasing the amount of water vapor in the air.

Hydraulic Gradient The change in static head per unit of distance in a given direction. If not specified, the direction generally is understood to be that of the maximum rate of decrease in head.

Hydrocarbon A compound composed of the elements carbon and hydrogen.

Hydrocyclone The basic form of most separators acting on the principle of centrifugal forces.

Hydrologic Cycle The continual circulation of water between the atmosphere, land, and sea by precipitation, evaporation, and transpiration.

Hydrolysis The chemical decomposition or splitting of a compound by reaction with water.

Hydrosphere The waters of Earth, including surface water, underground water, and water in the atmosphere.

Hydrostatic Head The height of a vertical column of water.

Igneous Rock A rock or mineral which solidified from molten matter that originated within the earth.

Imbiber Beads A commercial product used in contaminated water wells which will adsorb free oil, oil in solution, and a number of other petrochemicals.

Induced Infiltration Recharge to ground water by infiltration, either deliberate or inadvertent, from a body or surface water as a result of the ground water withdrawal and subsequent lowering of the ground water head below the surface-water level.

Induced Recharge *See* induced infiltration.

Infiltration The movement of water into a rock through its interstices or fractures or into the soil.

Interstitial Water Subsurface water located in the pore spaces of rocks.

Ion An atom or molecule possessing an electric charge.

Ion Exchange A reversible replacement of certain ions by others; the direction of the exchange depends upon the affinities of the ion exchanger for the ions present and the concentrations of the ions in the solution.

Ion-Exchange Capacity The quantity of a particular ion that can be replaced by ion exchange.

Ionic Bond The bond formed by the forces of electrostatic attraction of each group of atoms.

Isotopes Atoms with the same number of protons but a different number of neutrons. Isotopes of an element have essentially the same chemical properties.

Jetted Well A shallow-water well constructed by a high-velocity stream of water directed downward into the ground.

Landfill A general term indicating a disposal site of refuse, dirt from excavations, and junk.

Leachate Water that has percolated through soil, or a filter material, containing soluble substances and that, therefore, contains certain amounts of these substances in solution.

Leach Field Area where septic-tank effluent is distributed for natural leaching.

Lead-Base Fuel Any fuel using lead-containing chemical additives.

LeChâtelier's Principle If a stress is applied to a system at equilibrium, the equilibrium will shift to reduce the stress.

Legume Plants A family of plants including the peas, beans, clovers, and others, many of which are nitrogen fixing.

Lime The common name for calcium oxide (CaO).

Limestone A sedimentary rock consisting chiefly of calcium carbonate, primarily in the form of the mineral calcite.

Maul A heavy, long-handled hammer used to drive the well point in constructing a drive well.

Metamorphic Rock A rock that has been derived from preexisting rocks by mineralogical, chemical, and structural changes, essentially in the solid state, in response to marked changes in temperature, pressure, shearing stress, and chemical environment at depth in the earth's crust.

Meteoric Water Water derived from the atmosphere.

Methemoglobinemia A serious affliction of infants in which the oxygen-carrying capacity of hemoglobin is reduced as a result of a reaction with nitrite. Nitrite can be formed from nitrate by intestinal bacteria. Infant hemoglobin is replaced by adult hemoglobin, which is relatively immune to nitrite, by age of 6 months.

Micrometer A linear unit of measure equal to 0.00003937 in, abbreviated μm, and formerly called micron, $1\ \mu m = 1 \times 10^{-6}$ m.

Molecule The smallest particle of a substance capable of independent existence as a gas or as a definite entity in solution. Molecules may be either monatomic or polyatomic.

Montmorillonite A group of clay minerals which swell upon wetting and shrink upon drying.

Natural Softening The replacement of hardness-causing minerals by sodium and potassium by the normal flow of water in the ground.

Nematode Any roundworm of the phylum Nematoda, having unsegmented, threadlike bodies. Many are parasitic.

Neutralization The addition of either an acid to a base or a base to an acid to produce a neutral solution (usually considered to have a pH of 7).

Neutron A fundamental particle of the nucleus of an atom, having no charge and a mass equal to a proton.

Noncarbonate Hardness Water hardness due to the presence of compounds such as calcium and magnesium chlorides, sulfates, or nitrates; the excess of total hardness over total alkalinity.

Nucleus The central part of an atom containing all the protons and neutrons.

Organic Chemical A chemical characterized by its carbon-hydrogen structure.

Oxidation A chemical process in which electrons are removed from an atom, ion, or compound. Oxidation always occurs as part of an oxidation-reduction reaction.

Oxidizing Agents Any substance that oxidizes another substance and is itself reduced in the process.

Ozone An unstable form of oxygen (O_3), naturally occurring in the upper atmosphere, useful because of its strong oxidizing characteristics.

Pathogen Any organism which may cause disease.

Percentage Timer A timing device used in conjunction with chemical feeders to control the rate at which a chemical is added to the water system.

Percolation Field *See* leach field.

Permanent Hardness Water hardness due to the presence of noncarbonate hardness ions which will not be precipitated by boiling.

Permeability The capacity of a porous rock, sediment, or soil for transmitting a fluid without impairment of the structure of the medium. A measure of the relative ease of fluid flow under unequal pressure.

Pesticide Any chemical used for killing insects, weeds, rodents, or anything considered a pest.

pH The strength of the acid or base present measured on a scale that runs from 0 to 14. Technically the reciprocal of the logarithm of the hydrogen-ion concentration.

Photosynthesis The formation of carbohydrates in green plants by the chemical reaction of carbon dioxide and water in the presence of light and chlorophyll.

Phreatophyte A plant that obtains its water supply from the zone of saturation or through the capillary fringe and is characterized by a deep root system (i.e., willow and salt cedar).

Phytotoxin A poisonous plant constituent, physiologically harmful to humans and lower animals.

Piezometric Surface *See* potentiometric surface.

Pneumatic Hammer A device producing instantaneous pressures many times greater than normal by the abrupt acceleration or deceleration of water flow. Also called a *water hammer*.

Polychlorinated Biphenyl An extremely toxic chemical contained in transformers and capacitors.

Pore Water *See* interstitial water.

Postchlorination The application of chlorine to water following other water-treatment processes.

Potable Water suitable for human consumption.

Potentiometric Surface An imaginary surface representing the static head of ground water and defined by the level to which water will rise in a well.

Pothole Any pot-shaped pit or hole.

Prechlorination The application of chlorine to water prior to other water-treatment processes.

Precipitation (1) The condensation of water vapor in the atmosphere forming rain, snow, hail, or sleet. (2) The removal of a substance from a solution by increasing the concentration above saturation.

Primary Treatment The removal of floating, suspended, and settleable solids from untreated sewage.

Privy A pit toilet. Also called an *outhouse*.

Proton A fundamental particle of the nucleus of all atoms, carrying a unit positive charge of electricity and having a mass approximately equal to 1836 times that of an electron.

Protozoan Any of a large group of mostly microscopic, one-celled animals living chiefly in water.

Pseudomonas spp. A common group of sulfate-reducing bacteria causing accelerated corrosion of pipes.

Radical A group of two or more atoms that acts as a single atom and goes through a reaction unchanged, or is replaced by a single atom,

Radioactive Fallout Airborne particles of radioactive nuclei from nuclear explosions, which settle out of the atmosphere usually thousands of miles from the place of a detonation.

Radioactive Wastes Any radioactively contaminated or radioactive material which is of no practical use.

Reduction A chemical process in which electrons are added to an atom, ion, or compound. Reduction always occurs as part of an oxidation-reduction reaction.

Resin Tank The main body of a water softener or demineralizer containing the filter bed.

Reverse Osmosis A process for the removal of dissolved ions from water, in which water is forced through a semipermeable membrane, retaining most ions while transmitting the water.

Riser Pipe (1) A steel pipe used in jet drilling to carry water under pressure to the well point where the unconsolidated material is loosened. (2) The central pipe in a water conditioner which carries the softened water from the resin bed into the service lines.

Rotary Drilling A common drilling method that is a hydraulic process using a rotating drill pipe at the bottom of which is attached a hard-toothed drill bit. The rock chippings (cuttings) at the bottom of the hole are carried up by the circulation of a fluid (drilling mud) down through the drill pipe and forced up between the drill pipe and the well hole (annulus).

Rust A common name for iron oxide formed by the oxidation of iron in air in the presence of moisture, usually orange in color.

Sacrificial Anode An anode constructed of magnesium or other suitable metal and placed in a water-heater tank to protect the tank from corrosion.

Safe Drinking Water Act of 1974 The first major piece of United States water-management legislation that recognizes ground water as a fundamental component of the national water-resource network. The coverage of the Act is largely limited to the regulation of disposal wells that inject waste fluid underground and to the protection of aquifers designated by the EPA as "sole sources" of drinking water.

Saline Water or Solution Water or an aqueous solution containing an excessive amount of dissolved salts, usually over 10,000 mg/L.

Salt (1) A class of chemical compounds which can be formed by the neutralization of an acid with a base. (2) The common name for sodium chloride (table salt or common salt).

Screen The portion of a well casing that is slotted or perforated to permit the flow of water into the well.

Screening The operation of passing loose materials through a screen of known mesh so that constituent particles are separated into defined sizes.

Secondary Treatment The removal or reduction of suspended and dissolved solids and biological oxygen demand of effluent from primary treatment.

Sedimentary Rock A rock resulting from the consolidation of loose sediment that has accumulated from the transportation of older rocks or particles from the precipitation of chemicals in solution or from the remains or secretions of plants and animals.

Sedimentation The process of deposition of the solid suspended particles out of water, usually when the water has little or no movement.

Seep A natural discharge of water from the ground, not usually strong enough to sustain flow.

Separator A mechanical device to separate sand or silt from well water.

Septic Tank A common facility for the treatment of household sewage in rural areas.

Septum A porous surface in a diatomite filtering system, usually constructed of wire cloth or plastic-fiber cloth.

Sewage Domestic and industrial waste in a liquid or semiliquid state.

Silt A rock fragment or detrital particle smaller than a very fine sand grain and larger than coarse clay, having a diameter in the range of $^1/_{256}$ to $^1/_{16}$ mm (0.00016 to 0.0025 in).

Sinkhole A depression caused by the slumping or collapse of rock and soil due to a solution cavity. A typical feature of limestone terrain.

Slurry A very wet, highly mobile, semiviscous mixture or suspension of finely divided, insoluble matter.

Soap Curd An insoluble residue created by the reaction of sodium or potassium soaps and hardness ions (calcium and magnesium) in water.

Soda Ash The common name for sodium carbonate, a chemical compound used as an alkaline builder in some soap and detergent formulations, to neutralize acid water, and in the lime–soda ash water-treatment process.

Softening The process of removing the hardness-producing ions in exchange for less detrimental ions.

Soil Moisture Water in the soil either held by capillary attraction or in the process of movement toward the water table.

Solute A substance dissolved in a solution.

Solution A mixture of two or more substances in a single phase.

Solvent Any substance that can dissolve another substance.

Spontaneous Combustion The process of bursting into flame as a result of heat generated by internal chemical reaction.

Spring A place where ground water flows naturally from a rock or the soil onto the land surface or into a body of surface water.

Stockpile A pile or mound of raw material or product in storage or awaiting shipment.

Strainer A small valve used in many chemical-feed pumps to prevent particulate matter from entering the pump or system.

Submersible Pump A pump designed to fit into a well and operate below the water level.

Sulfate-Reducing Bacteria A group of bacteria capable of reducing sulfates in water to hydrogen sulfide gas. They have no sanitary significance and are classed as nuisance organisms.

Sump A pit, cistern, or other small containment structure used to collect or drain surface water.

Superchlorination The addition of large amounts of chlorine to a water supply to speed chemical reactions or ensure disinfection in a short contact time.

Surface Runoff Water flowing across the soil surface into a channel.

Synthetic Detergent A manufactured cleaning agent, such as linear alkyl sulfonate or alkyl benzene sulfonate.

Taste Threshold The minimum concentration of a chemical substance which can just be tasted.

Temporary Hardness Water hardness due to the presence of carbonate hardness ions which can be precipitated by heating the water.

Tertiary Treatment The removal or reduction of special chemicals and trace metals in effluent from secondary treatment.

Threshold-Odor Concentration The minimum concentration of an odor-causing substance which can just be detected.

Total Solids The weight of all solids, dissolved and suspended, organic and inorganic, per unit volume of water.

Trace Elements Any element present in minute quantities in an organism, soil, water, etc.

Tracer A substance used to determine the flow rate and direction of water movement in a stream or aquifer. Common salt is one of the most frequently used tracers.

Transpiration The discharge of water vapor from the leaf surface by plants; usually this water is absorbed by roots.

Trihalomethane Any of the chemical substances characterized by the halogen elements (i.e., fluorine, chlorine, bromine, etc.) attached to three positions on a methane molecule. These substances can be derived from a number of sources, are toxic in more than trace amounts, and reduce the germicidal activity of chlorine in treatment facilities when alkaline water is used.

Tritium A radioactive isotope of hydrogen containing two neutrons.

Turbidity A measure of the degree of opaqueness of water that is caused by the presence of suspended matter.

Ultraviolet Light Radiation from the wavelength range from 4000 Å, just beyond the violet region in the visible spectrum, to 40 Å on the border of the x-ray region.

U.S. Geological Survey The national governmental agency of the Department of the Interior created in 1849 to study and inform the public of the geology of the United States.

Valence The number of electrons in an atom which can easily be given up to bond or react with another group of atoms.

Venturi Throat A tube with a narrow throat that increases the velocity and decreases the pressure of the fluid passing through it. A venturi throat is commonly used to facilitate the introduction of a liquid or gas into flowing water. This is accomplished by joining the fluids just after the constriction in the tube where a vacuum is created, sucking the incoming fluid into the flowing water.

Wastepile Waste deliberately piled on the ground and abandoned or stored to await final disposal.

Water Bloom A prolific growth of plankton which may be so dense that it imparts a greenish, yellowish, or brownish color to the water.

Waterborne Disease A disease caused by a bacterium or organism able to live or be carried by water.

Water-Heater Banks The internal part of a water heater which is in direct contact with the water.

Water Table The surface between the zone of saturation and the zone of aeration; that surface of a body of unconfined ground water at which the pressure is equal to that of the atmosphere.

Well Head A reference to the well site, commonly to differentiate it from water quality or quantity at a different location in the water system.

Zeolite A group of hydrated sodium aluminosilicates, either natural or synthetic, with ion-exchange properties.

Zone of Weathering The superficial layer of Earth's crust above the water table that is subjected to the destructive agents of the atmosphere, and in which soils develop.

Index